A Scintillating GEM Detector for 2D Dose Imaging in Hadron Therapy

The research discussed in the present thesis has been performed in the section of Radiation, Detection and Matter of the department of Radiation, Radionuclides and Reactors, faculty of Applied Sciences, Delft University of Technology, Mekelweg 15, 2629JB Delft, The Netherlands.
This work is part of the research programme of the Stichting voor Fundamenteel Onderzoek der Materie (FOM), which is financially supported by the Nederlandse Organisatie voor Wetenschappelijk Onderzoek (NWO).

In the cover, designed by anNa, there is picture taken with the scintillating GEM detector in a 150 MeV proton beam with a collimator mask with 2 mm wide characters.

A Scintillating GEM Detector for 2D Dose Imaging in Hadron Therapy

Proefschrift

ter verkrijging van de graad van doctor
aan de Technische Universiteit Delft,
op gezag van de Rector Magnificus Prof. dr. ir. J.T. Fokkema,
voorzitter van het College voor Promoties,
in het openbaar te verdedigen op vrijdag 20 juni 2008 om 10.00 uur

door

Enrica SERAVALLI

Laurea di dottore in Fisica
Universiteit Padova, Italië
geboren te San Daniele, Udine, Italië

Dit proefschrift is goedgekeurd door de promotor:

Prof. dr. ir. C.W.E. van Eijk

Samenstelling promotiecommissie:

Rector Magnificus	voorzitter
Prof. dr. ir. C.W.E. van Eijk	Technische Universiteit Delft,promotor
Prof. dr. B.J.M. Heijmen	Erasmus MC-Daniel den Hoed Cancer Center, Rotterdam
Prof. dr. ir. T.H.J.J. van der Hagen	Delft University of Technology
Prof. dr. F. J. Beekman	Delft University of Technology
Dr. M. M. F. R. Fraga	LIP-Coimbra, Coimbra University, Portugal
Dr. ir. M. J. Schippers	Paul Scherrer Institut, Switzerland
Dr. ir. R. Kreuger	Delft University of Technology
Prof. dr. I.M. de Schepper	Delft University of Technology (reservelid)

ISBN 978-1-58603-883-0

Keywords: 2D dose imaging, scintillating gas, Gas Electron Multiplier, quality control of hadron beams

Published and distributed by IOS Press under the imprint Delft University Press

Publisher
IOS Press
Nieuwe Hemweg 6b
1013 BG Amsterdam
The Netherlands
tel: +31-20-688 3355
fax: +31-20-687 0019
email: info@iospress.nl
www.iospress.nl
www.dupress.nl

PRINTED IN THE NETHERLANDS

A Ettore e
mame, papi, Franci e Anna
che mi hanno continuamente
sostenuto durante questa
avventura

Un uomo si mette in marcia
per raggiungere, passo a passo, la saggezza.
Non è ancora arrivato.

Italo Calvino,Palomar

Contents

Introduction 1

Chapter 1 **Quality assurance in hadron therapy** 6

1.1 Radiotherapy with hadron beams 6
1.1.1 Role of radiotherapy 6
1.1.2 The ionization density of particle tracks 7
1.1.3 Hadron therapy 10
1.1.4 The planning of the treatment 11
1.1.5 Dose delivery 11
1.2. Quality assurance in radiotherapy 14
1.2.1 What is Quality Assurance? 14
1.2.2 Verification of dose delivery 15
1.2.3 Motivation 17
1.3 Thesis outline 22
1.4 References 23

Chapter 2 **Interaction of hadrons with matter** 26

2.1 Nature of the interaction of hadrons with matter 26
2.2 Energy loss 27
2.3 Range of charged particles 29
2.4 Energy and range straggling 30
2.5 Multiple coulomb scattering 31
2.6 References 32

Chapter 3 **Operation principles of the scintillating GEM detector** 33

3.1 The scintillating GEM detector 33
3.2 Gas amplification in a gas proportional counter 37
3.3 Gas Electron Multiplier 38
3.4 Light production in gas proportional scintillation (GPS) counter 42
3.5 Gas mixture considerations 43
3.6 Light scintillation in the scintillating double GEM detector 44
3.6.1 Ar/CF_4 emission spectrum 44
3.7 Light signal readout 48

3.8 The optical system 50
3.9 Electric signal readout 52
3.10 The gas flow system 54
3.11 References 54

Chapter 4 **Measuring procedures** 59

4.1 Irradiation setup 59
4.2 q_{out} definition 60
4.3 L_i definition 62
4.4 Picture background signals 62
4.4.1 The background component due to light emitted when GEMs are off 64
4.4.2 Residual background 68
4.5 Role of median filter 70
4.6 Error analysis 71
4.6.1 q_{out} uncertainty 71
4.6.2 L_i error analysis 73
4.6.2.1 CCD camera noise sources 73
4.6.2.2 Uncertainty on L_i 76
4.6.2.3 Considerations on the light signal magnitude 77
4.7 References 77

Chapter 5 **Scintillating GEM detector characterization in an x ray beam** 79

5.1 Experimental setup 79
5.2 Results and discussion 80
5.2.1 Scintillating GEM detector gain 81
5.2.2 Relation between L_i and q_{out} 82
5.2.3 "Start up effect" 83
5.2.4 Pressure dependence correction 85
5.2.5 Some Ar/CF_4 emission spectrum features 87
5.2.5.1 Ar/CF_4 emission spectrum as a function of GEMs voltage 87
5.2.5.2 Ar/CF_4 emission spectrum as a function of x ray tube current 88
5.2.6 GEM hole shape 89
5.2.7 Ar/CF_4 ratio 92
5.2.7.1 q_{out} and L_i 92

5.2.7.2 Emission spectra 93
5.2.8 Gas mixture purity 96
5.2.8.1 Varying the initial purity of the gas mixture 96
5.2.8.2 Closing the gas flow 97
5.2.8.3 Varying the gas flow rate 99
5.3 Conclusions 100
5.4 References 102

Chapter 6 **Scintillating GEM detector dosimetric properties in a proton beam** 104

6.1 Experimental setup 104
6.2 Nomenclature 106
6.3 Results and discussion 107
6.3.1 Light signal brightness 107
6.3.2 Response reproducibility 108
6.3.3 Dose linearity 108
6.3.4 Dose rate dependence 110
6.3.5 Spatial response 111
6.3.6 Time response 113
6.3.7 Pulsed beam feasibility study 114
6.3.8 Energy dependence 116
6.3.9 Field size dependence 121
6.3.10 Response uniformity 122
6.3.11 Radiation hardness 124
6.3.12 Tissue equivalence 124
6.4 Conclusions 125
6.5 References 127

Chapter 7 **First Results of a Scintillating GEM Detector for 2D dosimetry in an Alpha Beam** 128

7.1 Abstract 128
7.2 Introduction 128
7.3 Methods and Materials 131
7.3.1 The detector setup 131
7.3.2 The irradiation setup 132
7.3.3 Data acquisition and analysis 134
7.4 Results and Discussion 136
7.4.1 L_i and q_{out} 136

7.4.2 Charge transfer efficiency 138
7.4.3 Depth dose curves 139
7.4.4 Dose rate dependence 142
7.4.5 Ionization density 144
7.5 Conclusions 146
7.6 References 147

Chapter 8 **A scintillating gas detector for depth dose curve measurements in clinical carbon beams** 149

8.1 Abstract 149
8.2 Introduction 150
8.3 Methods and Materials 152
8.3.1 Detector setup 152
8.3.2 Irradiation setup 154
8.3.3 Data acquisition and analysis 155
8.3.4 The light signal 158
8.3.5 Uncertainties 159
8.4 Results and Discussion 160
8.4.1 Scintillating GEM detector response in a pulsed beam 160
8.4.2 Light intensity and spatial resolution 161
8.4.3 Depth dose curve 162
8.5 Conclusions 165
8.6 References 167

Chapter 9 **Conclusions and outlook** 169

Summary 178

Samenvatting 182

Acknowledgements 186

Curriculum Vitae 189

List of Publications 190

Introduction

The driving force in more than hundred-year history of radiotherapy has been the research for greater biological effectiveness and higher precision of the applied dose. The unavoidable doses given to healthy tissue have represented a limiting factor. The dose delivered to the tumour can be increased, by better targeting or in other words by better conforming the dose distribution to the target. Therefore, conformity has been the main goal of all recent developments in cancer radiotherapy. The rapid increase in hadrontherapy facilities can be seen in this context. Hadrontherapy is a radiotherapy technique that employs hadron beams, e.g. protons, and light nuclei such carbon. Charged particle beams show an increasing energy deposition with penetration distance leading to a maximum (the Bragg peak) near the end of the particle range. Behind this maximum, the energy deposition drops to zero within few mm. Therefore, the dose delivered by a hadron beam is well localized in depth with a small lateral spread allowing a precise scanning of the tumour volume. Thus, the dose deposited in the tumour can be increased and at the same time compared to photons, the integral dose in the healthy tissue can be reduced. Carbon ions have a higher LET (Linear Energy Transfer) than protons. They deposit a larger fraction of their energy at the end of their track, resulting in more intense local ionization that is considered highly effective against radiation-resistant tumours [1,2]. However, the fragmentation of the carbon ions produces a "tail" in the dose distribution behind the Bragg peak that implies irradiation of the immediately downstream tissues [3]. To date more than 50,000 patients have undergone proton treatment [4], and about 1000 patients were irradiated with carbon ions [5].

In radiotherapy, benefit for the patients can only be achieved if the treatment is delivered to them as planned. In particular, conformal treatments always bear the risk that an uncertainty in the delivered dose distribution may lead to an under dosage of the tumour, and/or over dosage outside the target volume. Therefore, the quantification of the dose that will be applied and verification that such dose will be delivered as planned are mandatory operations (also named quality control procedures) for reducing and avoiding treatment errors.

The dosimetric verification of planned dose distributions, prior to the patient treatment, is usually performed in a homogeneous water equivalent phantom (that simulates the biological tissues). The aim of such dosimetric verification is to check if the shape and the magnitude of a dose distribution in the phantom agree with the result of the treatment planning system[a]. In addition, it can be used to check the proper functioning of the beam delivery system or in other words the ability of the latter to deliver the desired dose distribution.

An ideal dosimeter for pre-treatment verification of dose distributions should be capable of acquiring in real time integrated tissue-equivalent signals in three dimensions with a fine spatial resolution ($\leq$ 1 mm). Moreover, it should have a linear response on a large dynamic range for high intensity radiation beams (~ 10^9 particles/ ($cm^2 \cdot s$)).

The conventional way of performing quality control of clinical beams is to measure the dose by means of a standard ionization chamber at many points in a water equivalent phantom. However, since the dose delivery varies in time, a measurement of the dose distribution in three dimensions is time consuming: for each point of measurement the full beam delivery sequence has to be repeated. To overcome this limitation, several methods are being developed to measure the dose in three or two dimensions. MRI gel dosimetry [b][6,7,8] provides 3D dose information but it has the disadvantage that a magnetic resonance imaging unit is needed for evaluation. Arrays of ionization chambers [9] present reliable dosimetric properties, but do not have satisfactory spatial resolution (~ 5-6 mm). Stacks of ionization chambers with strip-segmented anodes for 2D read out have a better spatial response but they do not provide full 2D dose information [10,11]. The use of stacks of films[c] [12] gives dose information with very high spatial resolution but to obtain digital data for analysis the film has first to be processed and scanned. Consequently, the film measurement evaluation is time consuming. Scintillating screens [13,14] coupled to a CCD camera allow online measurements of dose distributions with spatial resolution (~ 0.2 mm) as good as the film. However, the response of scintillating screens in high LET beams is affected by saturation.

---

[a] The dose distributions of the treatment planning system are recalculated in water in order to be compared with the ones measured in homogeneous phantoms.

[b] In MRI gel dosimetry, a humanoid phantom is irradiated according to the planned treatment of a patient. This results in a three-dimensional dose distribution. When gels are irradiated, free radicals created in the gel induce polymerisation restricted to the irradiated region. The number of free radicals, and thus the polymer yield, increases as a function of absorbed dose. The change in gel structure introduced by the polymerisation can be detected using for example MRI.

[c] Films are arranged between polyethylene absorbers to simulate an extended volume in which the applied dose distribution is verified with film detectors.

A patient treatment plan is usually composed by the superposition of beams having different energies. This gives additional complications as, for high LET radiation, the response of gels, films and scintillating screens depends on the energy. The response of these detectors decreases at low particle energies due to saturation. As a consequence, these detectors underestimate the dose at the Bragg peak depth[d] more than at the plateau of the curve. A correction for this energy dependence is difficult to apply because the composition of beam energies in the treatment plan at each position in the irradiated volume and the corresponding detector responses must be known. The aim of this work was to develop and characterize a 2D dosimetry system based on a scintillating gas detector for pre-treatment verification of dose distributions in hadron beams. The system is a follow up of the scintillating Gd_2O_2S:Tb ("Lanex") screen setup [13,15]. With a gas as primary detection medium, in high LET radiation beams a smaller energy dependence of the detector response is expected compared to the one of a Lanex screen because firstly, the light production process in a scintillating gas detector does not suffer from the quenching processes present in the Lanex screen. In fact, in the scintillating gas detector the photons are emitted by electron-excited gas molecules during the gas multiplication process. Secondly, the employed Ar/CF_4 scintillating gas mixture has better tissue equivalence[e] and a lower mass density than the scintillating screen. A spatial resolution comparable to that of a film is expected, and a faster and brighter response than that of a Lanex screen.
The dosimetry system we developed consists of a chamber filled with an Ar/CF_4 scintillating gas mixture at 1 atm, inside which two cascaded Gas Electron Multipliers (GEMs) are mounted. A GEM is a copper clad thin kapton foil with a regular pattern of sub mm holes [16]. The primary electrons, created in the detector's sensitive volume by the incoming beam, drift in an electric field towards the GEMs and undergo gas multiplication in the GEM holes. During this process, gas molecules are excited and subsequently they deexcite under fast light emission. Readout is done by means of a CCD camera. Since the amount of emitted light is proportional to the dose deposited by the incoming beam in the ionization chamber, the intensity distribution of the measured light spot is proportional to the 2D hadron dose distribution.

---

[d] A depth dose curve indicates the energy deposit of a hadron beam as a function of depth. It is characterized by a almost constant low entrance dose region, called plateau, followed by a high dose region, the Bragg peak.

[e] Tissue equivalent denotes a substance with absorbing and scattering properties for a given radiation that sufficiently match those of a certain biological tissue.

With such a setup positioned on the treatment couch before the patient treatment, one can get information about the shape of the energy deposit on the two coordinates in the plane perpendicular to the radiation beam. By placing a water bellows phantom in front of the detector, with respect to the beam direction, and by varying the water thickness in steps, from zero up to beyond the hadron range, a 3D dose distribution can be reconstructed. The latter can be then compared to the dose distribution computed in water by the treatment planning system.

References

[1] R.Orecchia et al., Particle Beam Therapy (Hadron therapy): Basis for Interest and Clinical Experience, European Journal of Cancer, 34, 1998, 459.
[2] U.Amaldi, G Kraft, Radiotherapy with beams of carbon ions, Rep. Prog. Phys., 68, 2005, 1861.
[3] P.Chauvel, Treatment planning with heavy ions, Radiat. Enviorn. Biophys., 34, 1995, 49.
[4] http://p-therapie.web.psi.ch/e/index.html.
[5] U.Amaldi, Hadrontherapy in the world, www.nupecc.org/iai2001/report/B32.pdf.
[6] U.Ramm et al., Three-dimensional BANGTM gel dosimetry in conformal carbon ion radiotherapy, Phys. Med. Biol., 45, 2000, N95.
[7] M.J.Maryansky et al., NMR relaxation enhancement in gels polymerized and cross-linked by ionizing radiation: A new approach to 3D dosimetry by MRI, Magn. Reson. Imaging, 11, 1993, 253.
[8] L.E.Olsson et al., MR imaging of absorbed dose distribution for radiotherapy using ferrous solphate gels, Phys. Med. Biol., 35, 1990, 1623.
[9] C.P.Karger et al., A system for three-dimensional dosimetric verification of treatment plans in intensity-modulated radiotherapy with heavy ions, Med. Phys., 26,1999, 2125.
[10] G.C.Bonazzola et al., Performances of a VLSI wide dynamic range current-to-frequency converter for strip ionization chambers, Nucl. Instr. and Meth., A 405, 1998, 111.
[11] C.Brusasco et al., A dosimetry system for fast measurement of 3D depth–dose profiles in charged-particle tumor therapy with scanning techniques, Nucl. Instr. and Meth., B 168, 2000, 578.
[12] A.J.Lomax et al., Intensity modulated proton therapy: a clinical example, Med. Phys., 28, 2001, 317.
[13] S.N.Boon et al., Fast 2D phantom dosimetry for scanning proton beams, Med. Phys., 25 1998, 464.

[14] S. Safai et al, Development of an inorganic scintillating mixture for proton beam verification dosimetry, Phys. Med. Biol., 49, 2004, 4637.
[15] S.N.Boon et al., Performance of a fluorescent screen and CCD camera as a two-dimensional dosimetry system for dynamic treatment techniques, Med. Phys, .27, 2000, 2198.
[16] F. Sauli, "GEM: A new concept for electron amplification in gas detectors," Nucl. Instr. and Meth., A 386, 1997,531.

Chapter 1

Quality assurance in radiotherapy

1.1 Radiotherapy with hadron beams

1.1.1 Role of radiotherapy

Cancer is a leading cause of death worldwide. In the European Union, it is estimated that in 2000 there were 1.892.000 incident cases of all forms of cancer (excluding non-melanoma skin cancers diagnosed) and that there were 1.156.000 deaths where cancer was the underlying cause [2]. Mortality from cancer in the world is projected to continue rising [1]. Treatment of malignant tumours is aimed at curing, prolonging and improving the quality of life of patients with cancer. Treatment may involve surgery, radiation therapy, chemotherapy, hormonal therapy, or some combination of these. As can been seen in Table 1.1, radiotherapy is involved in almost half of the curative treatments of loco-regional type (surgery and/or radiotherapy).

Modality	% of patients
Cured: surgery only	22
Cured: radiotherapy as prominent agent	12
Cured: surgery combined with radiotherapy	6
Cured: other systematic treatments	5
Not cured: uncontrolled primary tumour	18
Not cured: uncontrolled metastasis disease	37

Table 1.1 Cancer cures by treatment modality. Data taken from [3].

In particular, the goal of external radiation therapy [f] is the complete destruction of an entire tumour. Medical doctors plan treatment in order to

[f] External radiotherapy is a technique that uses radiation from a source outside the body, in contrast with internal radiotherapy in which radioactive sources are placed within the body in or near the tumor.

spare as much healthy tissue as possible. Often, this is difficult to achieve and complications arise due to irradiation of normal tissue. Therefore, a lot of efforts have been done in order to implement new techniques such as Intensity Modulated Radiation Therapy (IMRT) or Hadrontherapy, to increase the dose conformation to the tumour. IMRT makes use of several high energetic photon beams to irradiate the tumour from different directions. The intensity of the beams is varied across the irradiation field by means of variable collimators ("multi leaf-collimators") that are computer controlled. Hadrontherapy is a technique that employs hadron beams, e.g. protons, and light nuclei such carbon. Hadrons show an increasing energy deposition with penetration distance leading to a maximum (the Bragg peak) near the end of the particle range. Behind this maximum, the energy deposition drops to zero within few mm (section 1.1.3). Therefore, the dose delivered by a hadron beam is well localized in depth with a small lateral spread allowing a precise scanning of the tumour volume.

1.1.2 The ionization density of particle tracks

The amount of radiation is expressed in terms of absorbed dose, which is the deposited energy per unit mass. Its unit is Gray (Gy), 1 Gy = 1 J/Kg = $6.24 \cdot 10^{12}$ MeV/Kg.
Biological effects as a consequence of the absorption of energy from radiation may be caused by direct action of radiation: the target atoms can be ionized or excited, thus initiating a chain of events that leads to a biological change. Biological effects may also be caused by indirect action of radiation: the radiation interacts with water present in the cell to produce free radicals[g] that are able to diffuse far enough to reach and damage the critical target [4].
Equal doses of different types of radiation do not produce equal biological effects. The difference is due to the pattern of energy deposition at the microscopic level. The Relative Biological Effectiveness (RBE) is a measure of how damaging a given type of particle is when compared to an equivalent dose of x-rays. If a dose D of a given type of radiation produces a specific biological endpoint, then the RBE is defined as the ratio D_x/D, where D_x is the x ray dose needed under the same conditions to produce the same endpoint [4].

---

[g] Free radicals are atomic or molecular species with unpaired electrons. These unpaired electrons are usually highly reactive, so radicals are likely to take part in chemical reactions. Because of their reactivity, free radicals can participate in reactions resulting in cell damage.

The RBE depends on several parameters, among them the Linear Energy Transfer (LET). The LET is the average energy transferred per unit length of the track and it is usually expressed in keV/μm [4]. The LET is used to describe the density of ionization in particle tracks. Figure 1.1, shows examples of microdosimetric calculation of ionization tracks from gamma-rays and α-particles passing through a cell nucleus [5].

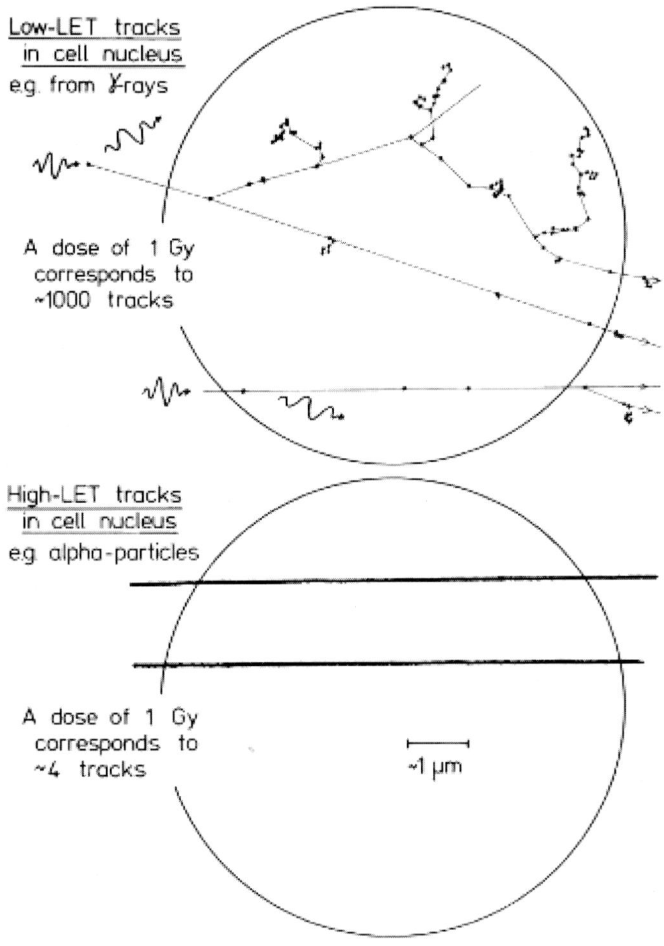

Figure 1.1 Schematic representation of a cell nucleus irradiated with two electron tracks from gamma rays (low LET) and two alpha particles tracks (high LET). Picture taken from [5] with permission from the author.

Gamma-rays, classified as low LET radiation, deposit much of their energy as single isolated ionizations or excitations. On the other hand, the high LET α-particles generate fewer tracks with more intense ionization along each track. In general, particles with LET values < 20 keV/μm are called low-LET, whereas those with larger LET values are categorized as high-LET particles. In Table 1.2, the LET values for the particle beams used in this work to irradiate the scintillating gas detector are listed. For each beam, the first listed energy corresponds to about the effective energy at the plateau of a depth dose curve[h], while the 2nd energy value is about the energy of that particular beam at the Bragg peak depth. Independently of the type of particle, the LET is higher at the Bragg peak depth energy. The LET of protons is overall lower than the LET of alpha particles and carbon ions.

	LET (keV/ μm) in liquid water
Protons	
100 MeV	0.70
30 MeV	1.838
Alpha particles	
200 MeV	4.983
40 MeV	18.22
^{12}C ions	
200 MeV/u [i]	16.36
60 MeV	37.21

Table 1.2 LET values for the particle beams used, in this work, to irradiate the scintillating gas detector. For each beam, the first listed energy corresponds to about the effective energy of the particle beam at the plateau of a depth dose curve, while the 2nd energy value is about the energy of that particular beam at the Bragg peak depth. Data taken from [8].

The efficiency of cell killing increases with LET, because of the increasing density of ionizations, excitations and free radicals produced in critical targets of the cell along the particle tracks. The radiobiological rationale for high-LET radiotherapy has mainly two reasons [6,7]: (1) cells can not repair the more extensive damage incurred by high LET radiation as easily as they can for low LET radiation damage. (2) Tumour cells are often hypoxic, i.e. they lack oxygen because of an inadequate supply of blood to the tumour. Such cells are more responsive to high LET than to low LET radiation. This

[h] A depth dose curve indicates the energy deposit of a hadron beam as a function of depth. It is characterized by a almost constant low entrance dose region, called plateau, followed by a high dose region, the Bragg peak.

[i] u is the atomic mass unit, that for carbon ions is equal to 12.

difference in response is due in part to the reduced production of oxidizing radicals under heavy hypoxic conditions for low LET radiation [6].

1.1.3 Hadron Therapy

External radiotherapy treatments are usually delivered with photon beams. Photon beams are characterized by a near-exponential decay of dose with the biological tissue depth, as shown, for example, for 6 MeV x rays in Figure 1.2. This particular depth dose dependence implies that structures in the entrance region received an equal or greater dose than the target volume situated at a certain tissue depth. Moreover, the dose is greater in the proximal region of the target volume than in the distal region. Three-dimensional Conformal Radiotherapy and, more recently, Intensity Modulated Radiation Therapy (IMRT) have been developed in order to improve the conformation of the dose distribution to the target volume [9,10].

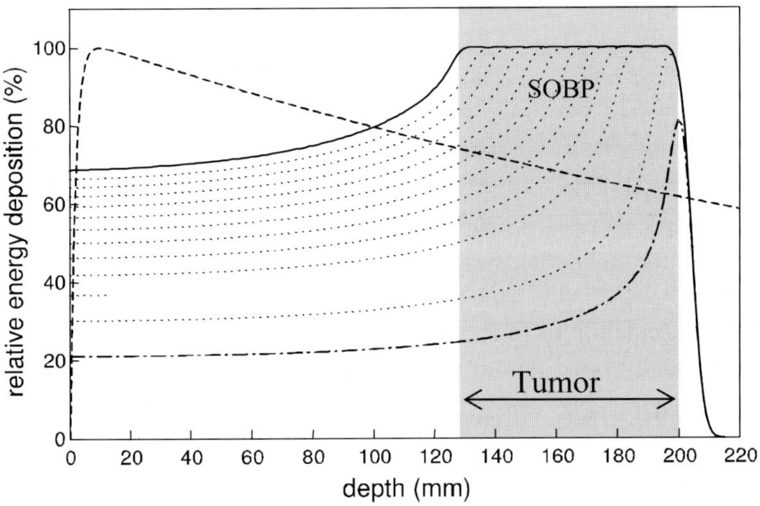

Figure 1.2 Comparison of depth dose curves of 6 MeV x rays (dashed line), 175 MeV protons showing the Bragg peak (dashed-dotted line), and energy modulated protons in water showing the spread out Bragg peak (SOBP) (solid line). The rectangular gray area indicates the irradiation target, tumour. Modified from [11].

The delivery of optimal dose conformation can also be achieved with charged particles, such protons and carbon ions. A monoenergetic hadron beam used for cancer therapy, e.g. a proton beam in Figure 1.2, is characterized by a depth-dose curve that has a low entrance dose (plateau), followed by a high-dose region (Bragg peak region) in the tumour area, and

followed by a steep fall-off to zero-dose distal to the target. Compared to photons, one can achieve with hadrons a considerable general reduction of the integral dose outside of the target volume. Because most of the tumours have a larger size than the Bragg peak width, hadron beams of different energies are combined to have an area of uniform ionization at depth sufficient to cover the intended target volume thickness (Spread Out Bragg Peak, SOBP). However, as can be seen in Figure 1.2, with a SOBP the ratio peak-dose / entrance-dose decreases and the SOBP entrance dose (skin dose) is higher than that of 6 MeV x-rays. Anyhow, for deep laying tumours this dose distribution is still very favourable compared to photons.
A disadvantage of hadron therapy is the large size and costs of the accelerator and beam lines needed for the transport of the beam.

1.1.4 The planning of the treatment

Many steps are involved in the external radiotherapy process. These steps can be divided into two different groups: preparation and treatment delivery (section 1.1.5). Often, the preparation starts with the production of immobilisation devices to accurately position and immobilise the patient at the treatment couch during daily treatments. Next, sequential computerized tomography (CT) slices are made of the patient in the treatment position to obtain 3D representation of the patient anatomy. To define a common point of reference between the actual anatomy and its 3D representation markers, drawn on the immobilisation device or patient skin, or implanted fiducials are used. Successively, the CT scans are transferred to the treatment planning system. The target volume and healthy tissue are then delineated on the CT slices and the dose to be delivered is prescribed. Based on the delineated structures and the dose prescription, a treatment plan is performed by optimising the number of beams, and beam energies. After approval of the resulting dose distribution by the radiation oncologist, the treatment preparation phase ends, and the patient may start with the treatment. Before starting the patient treatment, detailed dose verifications are made by means of dosimeters in water (section 1.2).

1.1.5 Dose Delivery

A hadron therapy facility consists of an accelerator (cyclotron or synchrotron), a beam transport and beam delivery lines. The latter can be "passive" or "active". In the "passive" delivery system, schematically represented in Figure 1.3, the hadron beam is scattered by material in the

beam in front of the patient in such a way as to produce a homogeneous flux of particles in the solid angle used for the irradiation. The dose is then shaped to the tumour in the lateral direction using collimators. A fast spinning wheel of variable thickness (range shifter wheel) introduces a variable amount of absorbing material in the beam as a function of time. The resulting modulation of the particle range can be chosen such as to produce a homogeneous region of the dose in depth (SOBP). Instead of the spinning wheel, a ridge filter can also be used for extending the dose in depth. A patient specific compensator bolus can be added to this set-up to shift the distal edge of the dose field so that the dose distribution is conformed more closely to the deepest side of the target volume. However, it is very difficult to reach a perfect congruence between the irradiated volume and the target. Therefore, a large fraction of proximal normal tissues is frequently contained in the high dose region.
All the necessary hardware must be adapted and in part created individually for each single dose field. This makes the beam delivery with multiple dose fields on a scattering gantry rather laborious [12,13].

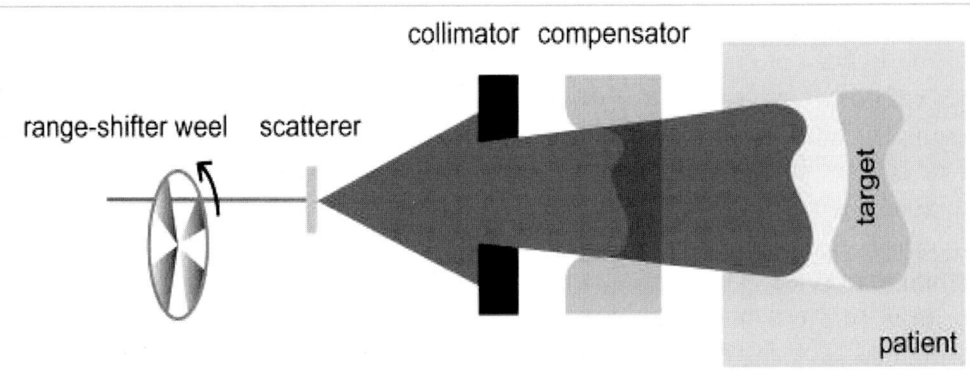

Figure 1.3 Schematic representation of a "passive" delivery system setup.

In an "active" delivery system, shown schematically in Figure 1.4, a mono-energetic Gaussian-shaped pencil beam (3-5 mm diameter) coming from the accelerator is scanned magnetically across the target volume. Typically, the beam is scanned in a zig-zag pattern in the x-y plane perpendicular to the beam direction. The depth scan (z) is done by means of energy variations. This method requires neither a collimator nor a compensator. The technique is based on the virtual dissection of the tumour in slices of equidistant particle ranges. One starts with the deepest layer (highest energy) and does one x-y scan. Successively, the energy is reduced and the next layer is painted, and so forth until all the layers have been delivered. A high conformity is achieved by changing the dosage and the position of each pencil beam individually.

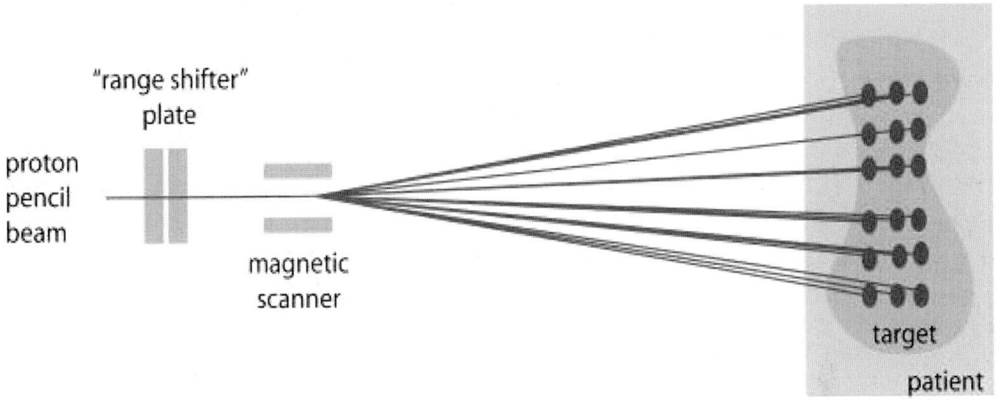

Figure 1.4 Schematic representation of an "active" delivery system showing sequentially delivered beam spots.

Figure 1.5 illustrates two examples of delivery patterns with a scanning beam. In the spot scanning (on the left of Figure 1.5), the predetermined dose is delivered to a given spot at a static position. Then, the beam is switched off and the magnet settings are changed to target the next spot. Dose is delivered to the next spot, and so on. This approach is practically implemented at the Paul Scherrer Institut (PSI), Switzerland for proton therapy. In the raster scanning (on the right of Figure 1.5), the beam is not switched off while it moves to the next position. This method is practically realized at the Gesellschaft für Schwerionenforschung (GSI), Germany for carbon ion treatment. One advantage of scanning with respect to a "passive" beam delivery system is that arbitrary shapes of uniform high-dose regions can be achieved with a single beam. A disadvantage is the higher sensitivity of this method to organ motion during scanning [12,13]. The dose distribution and homogeneity is in fact deformed by the motion of the tumour during the delivery.

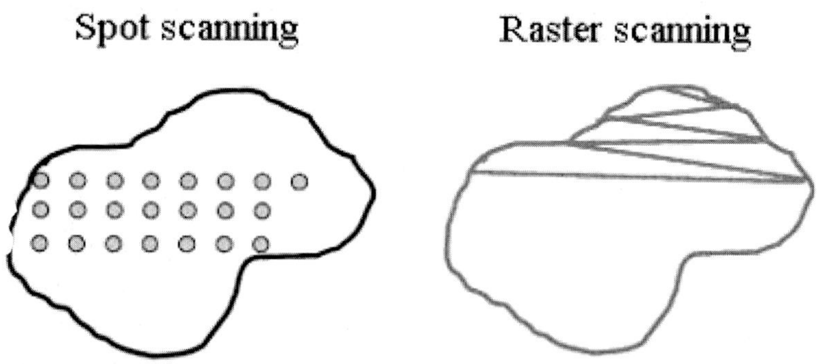

Figure 1.5 Examples of two delivery patterns for scanning beams. In the spot scanning delivery (on the left), the Gaussian shaped spots are delivered in discrete steps with a certain spot separation. The beam is switched off between the spots. In the raster scanning technique (on the right), the beam is switched on at the beginning of a scan line and remains on with constant intensity.

1.2 Quality assurance in radiotherapy

1.2.1 What is Quality Assurance?

Radiotherapy is a multidisciplinary specialty which uses complex equipment for treatment delivery. The success of the radiotherapy technique depends on how accurate the prescribed dose is delivered to the target volume. This implies that both the parameters related to the patient (diagnosis, indication for treatment, follow-up) and the procedures related to the technical aspects of providing the therapy (e.g. beam delivery system) should be subjected to careful quality control[j] in order to ensure consistency and safe fulfilling of the medical prescription. The required checks are grouped under the name of Quality Assurance (QA) [14]. Quality assurance of each radiotherapeutic treatment step is a prerequisite for the reduction of treatment uncertainties and the avoidance of errors [15, 16].
This thesis focuses on the development of a 2D dosimetry system for quality control of dose distributions prior to the patient irradiation.

[j] Quality control (ISO 9000) stands for the regulatory process through which (1) the actual quality performance is measured, (2) compared to existing standards and (3) finally the actions necessary to keep or regain conformance with the standards.

1.2.2 Verification of dose delivery

The pre-treatment verification of dose delivery aims at checking the proper transfer of treatment parameters to the treatment unit and the correct execution of the plan by the beam delivery system. The ultimate goal is to measure the dose distribution in three dimensions proving information concerning the behaviour of the dose laterally and in depth with a single measurement. Therefore, an ideal dosimeter[k] should be capable of acquiring in real time integrated tissue-equivalent signals in three dimensions with a fine spatial resolution ($\leq$ 1 mm). Moreover, it should have a linear response on a large dynamic range for high intensity radiation beams (~ 10^9 particles/($cm^2 \cdot s$)). The detector response should be LET independent and the detector should be radiation hard and water equivalent as much as possible [18].
The conventional way of performing quality control of clinical beams is to measure the dose by means of point dosimeter: the dosimeter is moved in three dimensions to cover the target volume and its response is recorded as a function of position. Examples of commonly used dosimeters for dose distribution verifications in hadron beams are:

- Thimble ionization chamber, in which a cylindrical anode is surrounded by graphite coated wall [19].
- Plane parallel ionization chamber, which consists of two parallel electrode foils separated by a small air gap (~ 2mm) [19];
- Diode detector, which consists of a p-type silicon junction [20,21];
- Diamond detector, this is a relatively new type of detector, the operation of which is based on the same principle as a diode [22].

However, verification of dose distributions by means of a single point detector is very time consuming. This applies especially to dynamically delivered dose distributions which vary with time (Figure 1.4): in order to verify the dose at a single point, the whole treatment field has to be applied completely. To overcome this limitation, two or three dimensional dosimetry systems, by means of which the dose can be measured in many points simultaneously, have been developed.
Examples of presently considered methods are:

---

[k] A dosimeter is a detector with a response that is proportional to the absorbed dose in a small region.

Two-dimensional:

- Ionization chambers array, which consists of two planes of several ionization chambers each. This dosimetry system provides a measurement of two, orthogonal dose profiles, at a single depth in water [23].

The disadvantage of detector arrays is that the spatial resolution (5-6 mm) is determined by the distance between single detectors along the array.

- Film dosimetry [24] provides 2D information about the dose distribution. The film can be stacked with a phantom material to measure 3D dose distributions [23]. This dosimetry system has a very good spatial resolution, but online evaluation is not possible.
- Scintillating screen coupled to a charge coupled device (CCD) camera [25,26] represents a follow up of the film dosimetry in the sense that it allows online evaluation of the dose distribution with spatial resolution (~ 0.2 mm) comparable to the film. However, the response of scintillating screens suffers from saturation.

Three-dimensional:

- Multiple ionization chambers each fixed in a hole of a polymethylmetacrylate mounting, which is attached to a motorized arm of a water phantom. Each ionization chamber can be positioned under visual control to any point in the phantom in order to measure any part of the 3D dose distribution [27].
- Stack of parallel plate ionization chambers sandwiched between phantom material plates to simulate the depth in the patient for the measurement of one dimensional depth dose profiles, integrated over the whole transversal area. The acquisition of the shape of the dose deposition transversal to the beam is performed by means of a multi-wire proportional counter (MWPC) coupled to a parallel plate ionization chamber and placed in front of the ionization chambers stack [28].
- MRI gel dosimetry [29,30] provides 3D dose information but with poor sensitivity (high irradiation doses are necessary). Moreover, a magnetic resonance imaging unit is needed for evaluation.

A patient treatment plan is usually composed by the superposition of beams having different energies. This gives additional complications as, for high LET radiation, the response of gels, films and scintillating screens depends on the energy. The response of these detectors decreases for low particle energies due to saturation. As a consequence, these detectors underestimate the dose at the Bragg peak depth with respect to the plateau of the depth dose curve. A correction for this energy dependence is not easy to perform. It could be applied only after determining the detector response for all possible energies in the treatment, having a priori knowledge of the composition of the particle field at each position in the irradiation volume.

1.2.3 Motivation

As already mentioned, for high LET radiation, the light emitted by the scintillating screens, e.g. $Gd_2O_2S:Tb$ ("Lanex"), does not show a linear response with the dose. For instance, in a measurement of a depth dose curve, this non linear response causes too small a signal at the Bragg peak depth. In Figure 1.6, a relative depth dose curve measured with the Lanex screen respectively in a proton beam and in a ^{12}C ion beam is compared to an air filled ionization chamber curve[1]. The curves are normalized to 1 at the minimum water depth.

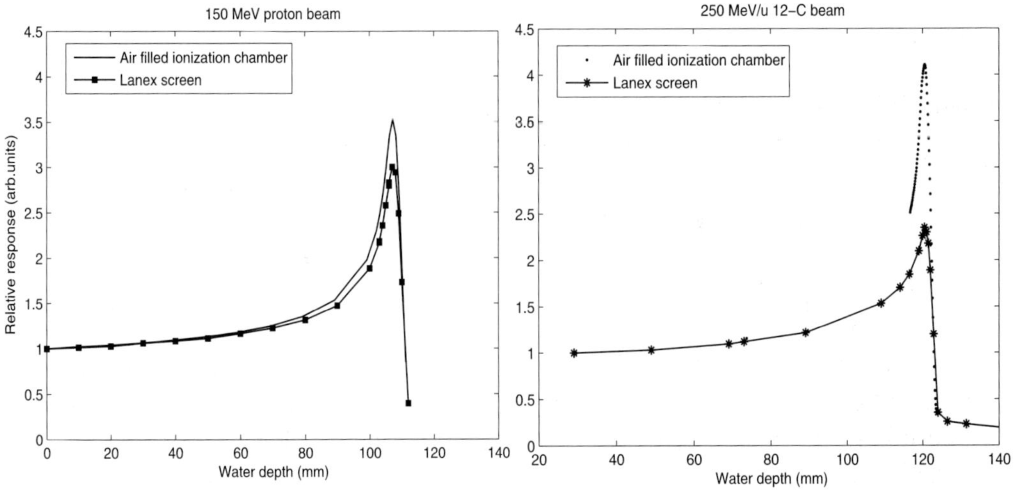

Figure 1.6 On the left, relative depth dose curve measured with the Lanex screen in a proton beam, compared to an air filled ionization chamber curve. On the right, relative depth dose curve measured with the Lanex screen in a ^{12}C ion beam, compared to an air filled ionization chamber curve. The curves are normalized to 1 at the minimum water phantom thickness.

In both cases, at the Bragg peak depth the Lanex screen signal is lower than

[1] Air filled ionization chambers are considered reference detectors in dosimetry.

the ionization chamber signal. The signal underestimation[m] is worse in a carbon ion beam, with higher LET than a proton beam (Table 1.2). The scintillating screen signal underestimation is due to a combination of effects: the quenching of the light production process, the non tissue equivalent screen composition, the averaging of the signal over the finite thickness of the screen [11].

The aim of this work was to develop and characterize a 2D dosimetry system based on a scintillating gas detector for pre treatment verification of dose distributions in hadron beams. With this dosimetry system positioned on the treatment couch before the patient treatment, one can get information about the shape of the energy deposit on the two coordinates in the plane perpendicular to the radiation beam. Moreover, by placing a water bellows phantom in front of the detector, with respect to the beam direction, and varying the water thickness in steps, from zero up to beyond the hadron range, a 3D dose distribution can be reconstructed. The latter can be then compared to the dose distribution computed in water by the treatment planning system.

The system is a follow up of the scintillating $Gd_2O_2S:Tb$ ("Lanex") screen dosimetry setup [11,25,26]. With a gas as primary detection medium, we expect a smaller energy dependence of the detector response compared to the scintillating Lanex screen signal, mainly because in a scintillating gas:

(1) The light production process is different than the one in a scintillating screen. Fluorescent or scintillating materials emit light due to excitation of states by the secondary electrons produced in the ionizing event, and subsequent radiative decay of specific energy levels in the material. Solid scintillators suffer from quenching when the LET of the particle track is high. Usually the model of Birks [32] is used to describe this process quantitatively. This model assumes that part of the scintillation centres along the ionization track is "damaged" or occupied in some way [33]. This part is proportional to *dE/dx*, the energy loss of the ionizing particle in the medium. A fraction of these centres dissipates the energy non-radiatively.

---

[m] The signal underestimation at the Bragg peak depth is calculated according to:

signal underestimation $= \dfrac{ppr_{Ic} - ppr_d}{ppr_{Ic}}$ where ppr_{Ic} is the peak to plateau ratio of the depth dose curve measured with the reference ionization chamber, while ppr_d is the peak to plateau ratio of a depth dose curve measured with a detector, the scintillating GEM detector or Lanex screen. The peak to plateau ratio is defined as the ratio of the signal measured at the Bragg peak depth and the one measured at the minimum water equivalent thickness when the depth dose curve is normalized to 1 at the minimum water equivalent thickness.

This leads to a quenching of the light yield. In a scintillating gas detector the light production process does not suffer from the quenching processes present in the Lanex screen because the photons are emitted by electron-excited gas molecules during the gas multiplication process (see chapter 3). However, the light production process could be affected by processes typical of gas proportional counters, such as gain non-uniformity, beam intensity, recombination, space charge and charge up effects.

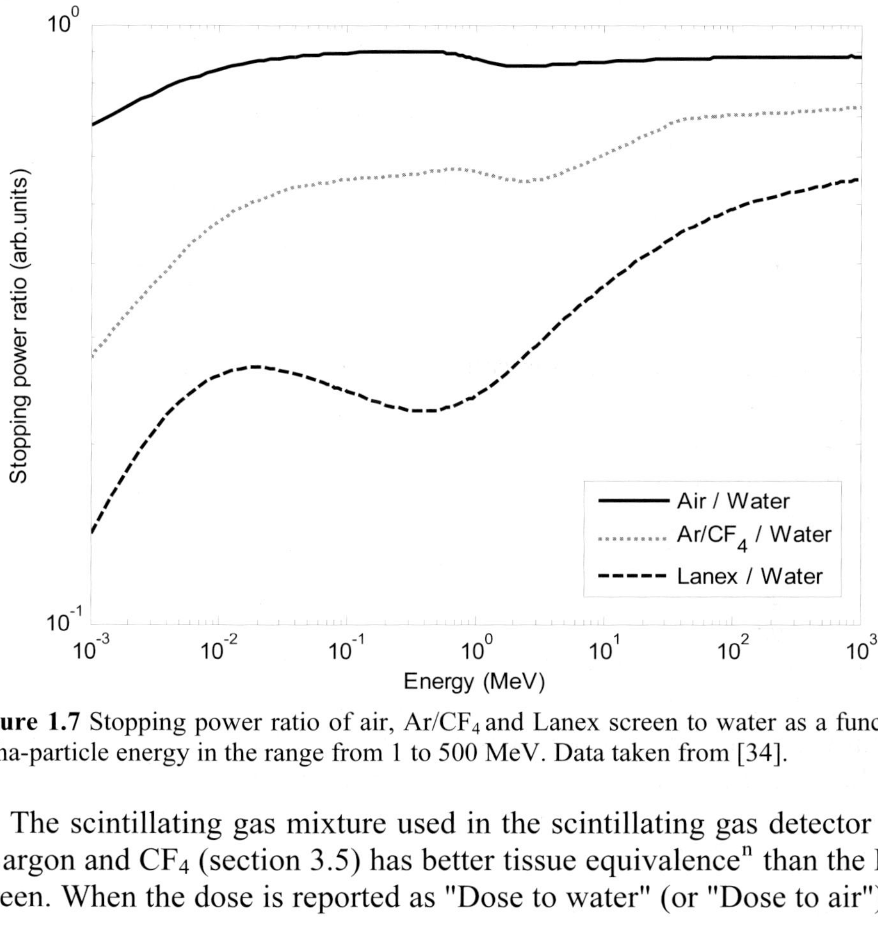

Figure 1.7 Stopping power ratio of air, Ar/CF_4 and Lanex screen to water as a function of alpha-particle energy in the range from 1 to 500 MeV. Data taken from [34].

(2) The scintillating gas mixture used in the scintillating gas detector based on argon and CF_4 (section 3.5) has better tissue equivalence[n] than the Lanex screen. When the dose is reported as "Dose to water" (or "Dose to air")[o], the

[n] Tissue equivalent denotes a substance with absorbing and scattering properties for a given radiation that sufficiently match those of a certain biological tissue.

[o] The main interest in dosimetry is the dose in the medium as it would be in absence of the detector. By means of the Bragg-Gray relation it is possible to relate the measured dose and the dose in the medium. $D_{med}=f \cdot D_{det}$, where D_{med} is the dose in the medium, D_{det} is the measured dose by means of the dosimeter and f is the medium-to-detector material stopping power ratio. Usually in dosimetry the medium is water.

ratio of the stopping power of the detector material relative to the stopping power in water (or air) has to be taken into account.
The non tissue equivalent detector composition implies that the ratio detector material-to-water stopping power is not one in the particle energy-range of interest for radiation therapy. Moreover, the stopping power ratio is not constant as, for example, the air-to-water stopping power ratio. In Figure 1.7, the stopping power ratio of air-, Ar/CF_4- and Lanex screen- to water are represented as a function of the alpha-particle energy in the range from 1 to 500 MeV. We consider, for example, a monoenergetic beam of 200 MeV alpha particles. Such a beam has energy at the Bragg peak depth of about 50 MeV. In Table 1.3, the stopping power ratio detector material-to-water calculated at the continuum, or plateau,[p] and at the Bragg peak depth of the depth dose curve is depicted for Lanex and Ar/CF_4. The calculated Ar/CF_4 signal underestimation at the Bragg peak depth due to the material properties with respect to the "ideal" water equivalent detector[q] is smaller than the one calculated for the Lanex screen.

	Alpha particle stopping power ratio detector-material / water			
	Bragg peak depth (50 MeV)	Plateau (200 MeV)	Ratio	Signal underestimation with respect to the "ideal" water equivalent detector
Lanex screen	0.461	0.516	0.893	1-0.893 = 0.11
Ar/CF_4 (92/08)[r]	0.705	0.715	0.986	1-0.986 = 0.01

Table 1.3 Alpha particle stopping power ratio of the detector material-to-water calculated at the plateau and at the Bragg peak depth for Lanex scintillating screen and Ar/CF_4 92/08. The signal underestimation with respect to the "ideal" water equivalent detector, detector for which the stopping power ratio detector material to water is one, is also reported. The stopping power data are from [34].

[p] The plateau of an alpha particle depth dose curve or Bragg curve is the initial low constant-dose region of the curve. The peak of the depth dose curve is the Bragg peak.
[q] An "ideal" water equivalent detector is a detector for which the stopping power ratio detector material / water is 1, independent of the particle energy.
[r] 92/08 indicates the ratio in volume percentage between the quantities of argon and CF_4 in the gas mixture.

(3) Ar/CF_4 has a lower density than the scintillating screen. Ideally, the dose D(x) at the detector position x can be calculated by:

$$D(x) = \frac{1}{\rho} \int \Phi(E,x) \cdot S(E) \cdot dE \tag{1.1}$$

where ρ is the detector material density, $\Phi(E,x)$ is the spectrum of particle energies E at position x and $S(E) = \frac{dE}{dx}$ the stopping power for particles with energy E. However, the detector is not infinitesimal thin which means that the stopping power has to be replaced by $\frac{\Delta E}{t}$ where ΔE is the mean energy loss in the detector with thickness t.

$$D(x) = \frac{1}{\rho} \int \Phi(E,x) \cdot \frac{\Delta E}{t} \cdot dE \tag{1.2}$$

This averaging can cause distortions in the observed Bragg peak, especially at low hadron energies for which the detector thickness *t* is not small anymore compared to the Bragg peak width [35].
The water equivalent thickness t_w of a gas detector is thinner than the one of a solid state detector because of the lower density, as can be seen in Table 1.4.

	ρ (g/cm^3)	t_w [s] (μm)
Lanex screen	5.88 [t]	353
Ar/CF_4 (92/08)	$1.827 \cdot 10^{-3}$	6

Table 1.4 Density and water equivalent thickness t_w for Lanex scintillating screen and Ar/CF_4 92/08. Lanex screen density from its data sheet; Ar and CF_4 density from respectively [34], and [36].

[s] The water equivalent thickness is calculated based on the following formula: $t_w = \frac{t_m \cdot \rho_m}{\rho_w}$
where t_w is the material water equivalent thickness; t_m is the material thickness; ρ_m and ρ_w are respectively the density of the material and liquid water.
[t] The density of Lanex is calculated taking into account the density of the Gd (5.3 g/cm^2) and the density of the binder (1.2 g/cm^2). The density of the acetate has not been taken into account, unlike S.Boon [31], because the acetate is not involved in the scintillation and so in the signal creation. The data are taken from the Kodak Lanex data sheet. The thickness of the Gd plus the binder is 60 μm.

Therefore, with a gaseous medium, the distortions at the Bragg peak due to the finite detector thickness are expected to be smaller than those expected for a solid state detector.
(4) Moreover, we expect the scintillating GEM detector having a spatial resolution comparable to the Lanex screen [37,38,39][u]; a faster response due to the fast drift of the electrons in the detector (< 1μs) and the short life times of the excited states [40,41].

Concluding, the objectives of this thesis are:

- to develop a scintillating Ar/CF_4 gas detector based on Gas Electron Multipliers [31];
- to understand its basic operation properties under x ray irradiation;
- to test the detector in charged particle beams in order to verify if it could be used as a 2D dosimeter for pre-treatment dose verifications. In particular, if it has smaller energy dependence than the Lanex scintillating screen.

1.3 Outline

In chapter 2, a basic discussion about the theory of the interaction of charged particles with matter and related subjects is presented.
In chapter 3, the scintillating GEM detector is described and its operation principles are explained.
In the chapter 4, the experimental setup and measuring procedures are described. The second part of the chapter deals with the data analysis: definition of the scintillating GEM detector outputs q_{out} and L_i; classification of background picture sources; uncertainty evaluation.
In chapter 5, the experiments performed to characterize the scintillating GEM detector under x ray irradiation are discussed.
The dosimetric properties of the scintillating GEM detector have been investigated in charged particles beams. In chapter 6, the experiments performed in a proton beam are discussed. The aim of these experiments was to verify if the scintillating GEM detector response is reproducible, uniform, linear with the dose and the field size, independent on the dose rate and LET. The energy dependence of the detector outputs, and the spatial and time response were also studied.

[u] Here, we are interested in the signal fall off when irradiating the detector with a step function shaped dose distribution (point spread function).

Chapter 7 illustrates measurements done with the scintillating GEM detector in an alpha beam. In chapter 8, the first results of the use of the scintillating GEM detector as a dosimeter in a clinical carbon ions beam are presented. For clarity, the measurements in hadron beams have been described in this order, but actually the experiments in the alpha particle beam were performed earlier than the proton experiments.
Chapter 9 contains the conclusions of this work and the suggested directions for further research.

1.4 References

[1] European code against cancer, www.cancercode.org.
[2] World Health Organization, www.who.
[3] Annex A, Europe Against Cancer, IOS, 1994
[4] E.Hall, Radiobiology for the Radiologist, 5th edition Lippincott Williams & Wilkins, 2000
[5] D.T.Goodhead, Spatial and Temporal Distribution of Energy, *Health Physics*, 55, 1988, 231.
[6] P. L. Petti, A.J.Lennox, Hadronic Therapy, *Annu. Rev. Nucl. Part. Sci*, 44, 1994, 155; [7] U.Amaldi, G.Kraft, Radiotherapy with beams of carbon ions, *Rep. Prog. Phys.*, 68, 2005, 1861
[8] D.E.Watt, Quantities for Dosimetry of ionizing radiations in liquid water, Taylor and Francis, 1996
[9] S.Webb, The physics of Three-Dimensional Radiation Therapy, Insitute of Physics Publishing, Bristol, 1993.
[10] S.Webb, Intensity modulated radiation therapy, Insitute of Physics Publishing, Bristol.
[11] S.Boon, "Dosimetry and quality control of scanning proton beams," PhD thesis, University of Groningen, The Netherlands, 1998.
[12] W.Chu, et al., Instrumentation for treatment of cancer using protons and light ions beams, *Rev. Sci. Instrum.*, 64, 1993, 2005.
[13] E. Pedroni, Latest development in proton therapy, In Procceedings of the 7th European Particle Accelerator Conference, *EPAC 2000*, Vienna 2000, 240.
[14] WHO, Quality assurance in Radiotherapy, Geneva, 1988
[15] B.J. Mijnheer, Quality assurance in radiotherapy:physical and technical aspects. *Quality Assurance in Health Care*, 4, 2002, 9.
[16] T. K. Yeung, Quality assurance in radiotherapy:evaluation of errors and incidents recorded over a 10 year period, *Radiotherapy and Oncology*, 74, 2005, 283.

[17] *L. Raffaele e al.,* Proton beam dosimetry for the CATANA Project, *Physica Medica*, XVII, Supplement 3, 2001, 35.
[18] M. Bucciolini et al., Validation protocols for dosimeters in highly conformal radiotherapy, Methods and Advanced Equipment for Simulation and treatment in Radiation Oncology (MAESTRO) project.
[19] J.Medin, Ionization chamber dosimetry of proton beams using cylindrical and plane parallel chamber. Nw versus Nk ion chamber calibrations, *Phys. Med. Biol.* 40, 1995, 1161.
[20] G.Rikher, Patient dose measurements in photon fields by means of silicon semiconductor detectors, *Med. Phys*., 14, 1987, 870.
[21] T. C. Zhu et al., Performance evaluation of a diode array for enhanced dynamic wedge dosimetry, *Med. Phys.*, 24, 1997, 1173.
[22] S.M.Vatnisky, Application of solid state detectors for dosimetry in clinical proton beams, *Med. Phys.* 22, 1995, 469
[23] A.J. Lomax, Treatment planning and verification of proton therapy using spot scanning: Initial experiences, *Med. Phys.,* 31, 2004, 3150
[24] L.J.van Battum, Film dosimetry of clinical electron beams, *Int. J. Rad. Oncol. Biol. Phys.* 18, 1990, 69.
[25] S.N.Boon et al.,_Fast 2D phantom dosimetry for scanning proton beams, *Med. Phys.,* 25 1998, 464.
[26] S.N.Boon et al.,_Performance of a fluorescent screen and CCD camera as a two-dimensional dosimetry system for dynamic treatment techniques, *Med. Phys.,* 27, 2000, 2198.
[27] C.P.Karger et al., A system for three-dimensional dosimetric verification of treatment plans in intensity-modulated radiotherapy with heavy ions, *Med. Phys*., 26,1999, 2125.
[28] C.Brusasco et al., A dosimetry system for fast measurement of 3D depth–dose profiles in charged-particle tumour therapy with scanning techniques, *Nucl. Instrum and Methods, B* 168, 2000, 578.
[29] U.Ramm et al., Three-dimensional BANG^{TM} gel dosimetry in conformal carbon ion radiotherapy, *Phys. Med. Biol*., 45, 2000, N95.
[30] M.J.Maryansky et al., NMR relaxation enhancement in gels polymerized and cross-linked by ionizing radiation: A new approach to 3D dosimetry by MRI, *Magn. Reson. Imaging,* 11, 1993, 253.
[31] F. Sauli, "GEM: A new concept for electron amplification in gas detectors," *Nucl. Instrum and Methods, A* 386, 1997,531.
[32] J.B.Birks, *The Theory and Practice of Scintillation Counting*, Pergamon, New York, 1967.
[33] R.B.Murray, A.Meyer, "Scintillating Response of Activated Inorganic crystals to Various Charged Particles," *Phys. Rev*., 122, 1961, 815.

[34] http://physics.nist.gov/PhysRefData/Star/Text/contents.html
[35] Bichsel, Calculated Bragg curves for ionization chambers of different shapes, *Med. Phys*., 22, 1995, 1721.
[36] http://encyclopedia.airliquide.com/Encyclopedia.asp?GasID=61;
[37] F.Fraga et al., Performance of a Tracking Device Based on the GEM Scintillation, *IEEE Transactions on Nuclear Physics*, 49, 2002, 281.
[38] S.Bachmaan et al., High rate X-ray imaging using multi-GEM detectors with a novel readout design, *Nucl. Instrum and Methods,* A 478, 2002,104.
[39] G.P.Guedes etal., Two-dimensional GEM imaging detector with delay-line readout, *Nucl. Instrum. and Methods,* A 513, 2003, 473.
[40] M.M.R.Fraga et al., Pressure Dependence of Secondary NIR Scintillation in Ar and Ar/CF_4, *IEEE Transactions on Nuclear Physics* ,48, 2001, 330.
[41] J.E.Hesser, K.Dressler, Radiative Lifetimes of Ultraviolet Emission Systems Excited in BF_3, CF_4, and SiF_4*, *The Journal of Chemical Physics*, 47, 1967, 3443.

Chapter 2

Interaction of hadrons with matter

Hadrons are subatomic particles subject to the strong nuclear force that binds particles together within the atomic nucleus. The name hadron is derived from a Greek word meaning “strong”. Typical hadrons are protons and neutrons that make up atomic nuclei and, by extension, those nuclei themselves. The hadrons currently employed in radiotherapy are neutrons, protons and the nuclei of light atoms such as helium, carbon, oxygen and neon (without, or with, some of their bound electrons); the latter are generally referred to as light ions.
This chapter focuses on the way hadrons interact with matter. Only topics relevant for the understanding of the work described in this thesis are discussed.

2.1 Nature of the interaction of hadrons with matter

Charged particles interact with matter primarily through the Coulomb force between their positive charge and the negative charge of the orbital electrons within the absorber atoms.
When a charged particle enters any absorbing medium, it interacts immediately with many electrons. In any of such encounters, the electron feels an impulse from the attractive Coulomb force as the particle passes in its vicinity. Depending on the proximity of the encounter, this impulse may be sufficient either to raise the electron to a higher-lying shell within the absorber atom (excitation) or to remove completely the electron from the atom (ionization). As a consequence of this encounter, part of the particle energy is transferred to the electron, and therefore, the particle velocity decreases. The maximum energy that can be transferred to the electron in a single collision is a small fraction, about 1/500, of the particle energy per nucleon. The charged particle deflection during the collision is negligible. Therefore, charged particles travel an almost straight path through the matter, losing energy almost continuously through many collisions with

atomic electrons. The products of these encounters in the absorber are excited atoms or ion pairs. The W-value is defined as the average energy lost by the incident particle per ion pair formed. Due to the competing mechanism of energy loss, i.e. excitation, the W-value is always greater than the ionization energy.
In some collisions, an electron may gain enough energy so that, after having left its parent atom, it still creates further ions. These energetic electrons are called delta rays.

2.2 Energy loss

The stopping power S is the mean energy loss per unit path length in a material, dE/dx.
dE/dx is described by the Bethe Bloch formula:

$$S = -\frac{dE}{dx} = 2\pi N_a r_e^2 m_e c^2 \rho \frac{Z}{A}\frac{z^2}{\beta^2}\left[\ln\left(\frac{2m_e c^2\beta^2 T_{\max}}{I_0^2(1-\beta^2)}\right) - 2\beta^2\right] \quad (2.1)$$

where N_a is the Avogadro's number; r_e is classical electron radius; $m_e c^2$ electron rest mass energy; β is the ratio of the incident particle velocity to the velocity of the light; ρ is the density of the absorbing material; Z is the atomic number of the absorbing material; A is the atomic weight of absorbing material; z is the charge of incident particle in units of e; T_{max} is the largest possible energy loss of the particle in a single collision with an electron; I_0 is the excitation potential of the material;
In first approximation, for a non relativist particle:

- $$S = -\frac{dE}{dx} \propto \frac{1}{\beta^2} \quad (2.2)$$

Particles with low velocity, and so low energy, have higher stopping power than that of fastest particles.

- for different charged particles with the same velocity, $S = -\frac{dE}{dx} \propto z^2$ (2.3)

  Particles with the highest charge will have the largest stopping power.

- for different absorber materials, $S = -\frac{dE}{dx} \propto \rho$ (2.4)

  High density materials will result in the largest stopping power.

Commonly, the hadron stopping power tables report the mass stopping power expressed in $\mathrm{MeVcm^2g^{-1}}$, which is obtained by dividing the stopping power by the density of the material, ρ. In Figure 2.1 the stopping power in liquid water of protons, alpha particles and carbon ions is graphed as a function of the energy. Going from high to low energies, the stopping power increases in the high energy region due to inverse square dependence on the particle velocity (formula 2.2). However, for lower energies when the logarithmic term in formula 2.1 starts to dominate, a peak occurs. The linear rate of energy loss is maximal there. Carbon ions have the largest stopping power compared to protons and alpha particles because they have higher charge (formula 2.3)

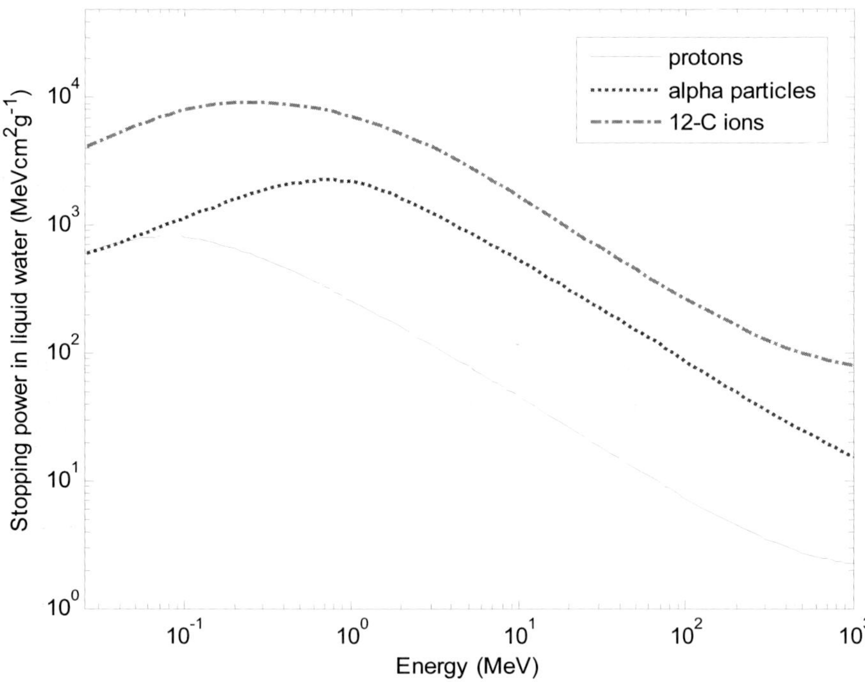

Figure 2.1 Stopping power in liquid water of protons, alpha particles and carbon ions graphed as a function of the particle energy. Data taken from NIST [2].

2.3 Range of charged particles

The range R of a charged particle in a particular absorbing medium is an experimental concept providing the thickness of an absorber that the particle can just penetrate. It depends on the particle's kinetic energy, mass as well as charge, and on the absorbing medium composition. Various definitions of range that depend on the method employed in the range determination are in common use [3].
The so called *continuous slowing down approximation range* (CSDA) is the average distance in a medium travelled by the charged particle of a specific energy E from the entrance point to the point where the energy is nearly zero.
It can be determined by integrating the reciprocal of the stopping power from 0 to E, assuming that the charged particles lose their energy continuously along their tracks at a rate given by the stopping power.

$$R = \int_0^E -\left(\frac{dE}{dx}\right)^{-1} dE \qquad (2.5)$$

where R is the charge particle range.
The CSDA range is a very good approximation to the average range of the particle because of the essential rectilinear path of the charged particles in the absorber.
In Figure 2.2, the range in water for proton, alpha particles and carbon ions is shown as a function of the particle energy.
In Table 2.1, the range values in water for the particles and the energies used to irradiate the scintillating gas detector are reported. Carbon ions have about the same range in water of the protons but for an energy that is about twenty four times higher than that of protons.

	Energy (MeV)	CSDA range in liquid water (μm)
Proton	100	$8.05 \cdot 10^4$
Alpha particle	200	$2.23 \cdot 10^4$
^{12}C	200 /u	$8.88 \cdot 10^4$

Table 2.1 Range values for the particles used for irradiating the scintillating gas detector. The range values are from [4].

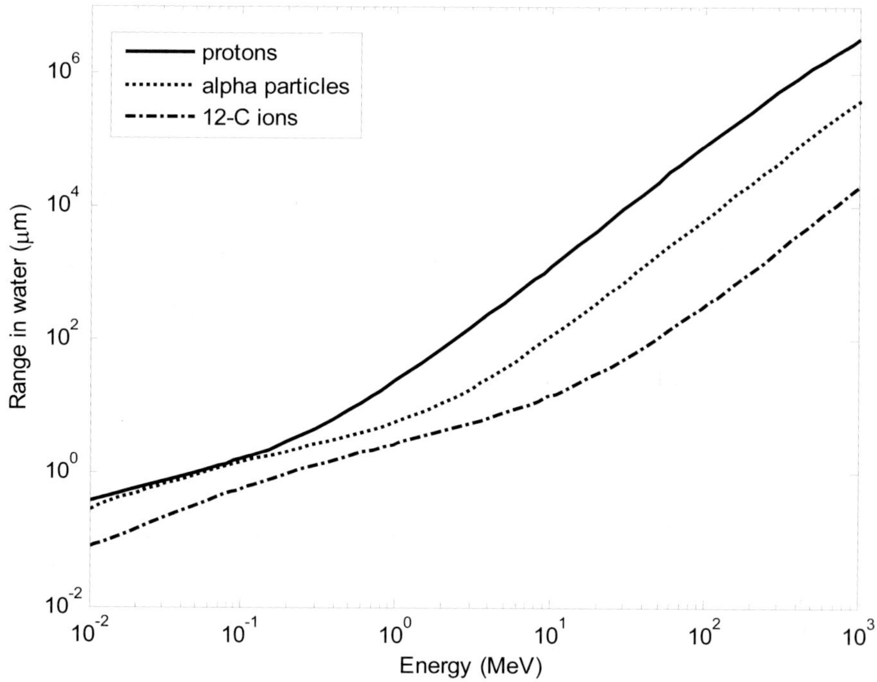

Figure 2.2 The range in water for proton, alpha particles and carbon ions as a function of the particle energy. Data are from [4].

2.4 Energy and range straggling

As a charged particle penetrates matter, statistical fluctuations occur in the number of collisions along its path and in the amount of energy lost in each collision. As a result, a number of identical particles starting out under identical conditions will show a distribution of energies (energy straggling) as they pass a given depth and a distribution of path-lengths (range straggling) traversed before they stop. The energy transferred to the atomic electrons in each collision with the charged particle is a stochastic quantity characterized by a Poisson-like distribution. In fact, in most of the hadron-electron collisions only a small amount of energy is transferred from the charged particle to the electron, due to the large ratio of hadron and electron mass. In a thick layer, t, a large number of collisions occur and the energy loss is expected to be distributed according to a Gaussian with a width σ_E given by [5]:

$$\sigma_E^2 = ktT_{\max}\left(1-\frac{\beta^2}{2}\right) \tag{2.6}$$

$$\text{with } k = 2\pi r_e^2 m_e c^2 N_a z^2 \left(\frac{Z}{A}\right)\frac{1}{\beta^2} \tag{2.7}$$

The Gaussian approximation is valid if the hadron energy can be assumed to be constant during the passage through the absorber.
In first approximation, the range straggling is also distributed according to a Gaussian with a width σ_R [5]. Range straggling decreases quadratically with the atomic number of the projectile ion [6].
The energy loss straggling contributes to the increase in dose in front of and beyond the Bragg peak.
Range straggling has large effects on the dose fall-off gradient beyond the Bragg peak of a particle traversing matter, reducing the height and increasing the width of the peak. The broadening of the Bragg peak is of prime importance when treating a tumour close to an organ at risk in front of or distal to the tumour [7]. The Bragg peak of carbon ion beams is narrower than that of protons because for the former the range straggling is smaller than that of protons.

2.5 Multiple coulomb scattering

Multiple coulomb scattering refers to the deflection of a charged particle from the original direction after every coulomb interaction. The scattering angle is the angle between the original direction of the particle and the final direction after the interaction.
Hadrons experience a deflection when they pass in the neighbourhood of a nucleus as a result of the combined interaction of the coulomb and hadronic field of the nucleus (the deflection caused by collisions with electrons can be neglected because of the mass ratio).
A detailed discussion about the multiple scattering theory can be found in [5].
Due to the multiple scattering, a pencil beam of charged particles is spread into a diverging beam as it penetrates a target. The magnitude of the spreading increases with the atomic number of the material. The increased angular spread extends the size of the penumbra and influences the spatial distribution of the Bragg peak [7].

Due to the mass difference, the broadening of a carbon ion beam as a function of water depth is much less than that of a proton beam, as can we see for example in Figure 3 of [6]. Thus, deep-seated tumours can be irradiated more precisely with carbon ions.

2.6 References

[1] G.F.Knoll, Radiation Detection and Measurement, 2nd Edition, John Wiley and Sons.
[2] http://physics.nist.gov/PhysRefData/Star/Text/contents.html
[3] E.B.Podgorsak, Radiation physics for medical physicists, Springer, 2006
[4] D.E.Watt, Quantities for Dosimetry of ionizing radiations in liquid water, Taylor and Francis, 1996
[5] H.A.Bethe and J.Ashkin, Passage of radiation through matter, Experimental nuclear physics, E. Segre' edition, vol I, Wiley, 1953
[6] W. K. Weyrather, J. Debus, Particle Beams for Cancer Therapy, *Clinical Oncology,* 15, 2003, S23–S28.
[7] M Hollmark et al., Influence of multiple scattering and energy loss straggling on the absorbed dose distributions of therapeutic light ion beams: I. Analytical pencil beam model, *Phys. Med. Biol.*, 49, 2004, 3247.

Chapter 3

Operation principles of the scintillating GEM detector

As mentioned in the Introduction and Chapter 1, a 2D dosimetry system, named scintillating GEM detector, based on a scintillating Ar/CF_4 gas mixture and on Gas Electron Multipliers (GEMs) and has been developed. A GEM is a copper clad think Kapton foil perforated by a regular pattern of sub mm holes. In the GEM holes gas multiplication takes place.
The scintillating GEM detector is a combination of a double GEM detector and a gas proportional scintillation counter (GPS). In fact, the production of photons in the detector is related to the same processes that occur in gas proportional scintillation counters.
The measurements reported in Chapter 7, which were performed first, were carried out with a detector already existing (section 7.3.1). Since this detector had some limitations, such as the too small entrance window (Chapter 7), a new detector was developed and used for the measurements discussed in Chapter 5, 6, and 8. This latest detector is in this chapter described together with its operation principles.

3.1 The scintillating GEM detector

A schematic representation of the scintillating GEM detector is given in Figure 3.1, a three dimensional drawing is represented in Figure 3.2 and detector photographs are shown in Figure 3.3. The detector consists of a aluminium chamber (350×350×50 mm^3) flushed with an Ar/CF_4 scintillating gas mixture at 1 atm. The 150×150 mm^2 entrance window is made of a 25 μm thick aluminized Mylar foil[V]. The cathode consists of

[V] The entrance window is made by an aluminized Mylar foil because a light tight material is needed as entrance window in order to prevent ambient light to enter the detector. Moreover, the entrance window must be made by a material that is at the same time not-easy to break and thin (to degrade the beam energy as less as possible).

160×160 mm^2 25 μm thick aluminized Mylar foil. The Al Mylar foil is glued, with Bondmaster two components Expoxy, onto an Al frame and it is located 0.5 mm downstream of the entrance window, with respect to the radiation beam direction. Two cascaded 100×100 mm^2 Gas Electron Multipliers (GEMs), produced at CERN [1,2], and named respectively GEM_1 and GEM_2, have been mounted. The GEMs used in this work have 80 μm (big holes) or 60 μm (small holes) diameter double conical holes with a pitch of respectively 140 μm and 90 μm, and are glued onto Al frames.

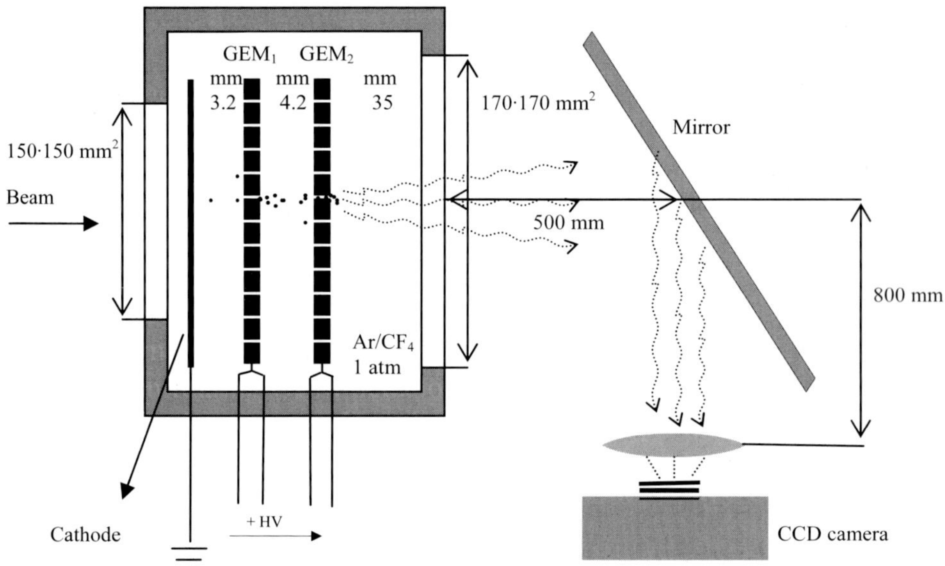

Figure 3.1 Schematic representation of the scintillating GEM detector not to scale.

The gap between the cathode and GEM_1 (*drift gap*) is 3.2 mm, while the gap between the two GEMs (*transfer gap*) is 4.2 mm. The 170×170 mm^2 exit window is made of 3 mm-thick Duran 50 glass and it is located 35 mm behind GEM_2. The exit window is made by glass in order to allow the transmittance of the photons produced by the scintillating gas mixture (section 3.6). In Figure 4.4, the transmission curve of the Duran 50 glass is shown together with the 92/08 Ar/CF_4 emission spectrum. As can been seen, the Duran 50 glass transmission is constant in the Ar/CF_4 emission spectrum wavelength range. The gap between the surface of GEM_2 facing the exit window and the exit window itself is named *light gap*. The spacers between the cathode and the GEMs that are shown in Figure 3.2 are made of aluminium oxide.

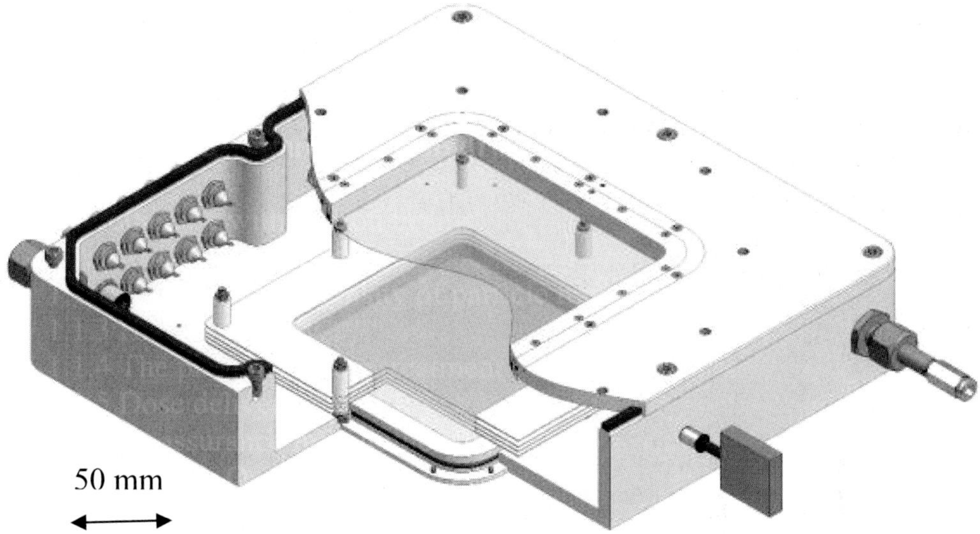

Figure 3.2 3D design of the scintillating GEM detector chamber. The entrance window is facing down. Frames holding the cathode foil next to the entrance window and the GEM foils are shown at the bottom. The square dark gray box on the right represents the pressure meter. On the same chamber side, the gas outlet is shown.

Eight pins, on which the cathode and GEM frames are mounted, are made of a metallic pivot covered by a ceramic cylinder. The Al frames, on which the GEMs are glued, are grounded to the detector box. A pressure sensor (Motorola MPX4115AS, case 867E) visible in Figure 3.2 is used to monitor the gas pressure variations inside the detector, while the gas internal temperature is controlled by means of a temperature sensor [3].
In presence of proper electric fields in the chamber gaps and across the GEMs, the primary electrons created in the drift gap (sensitive volume) by the incoming radiation beam drift towards the GEM_1 holes (section 3.3). In the GEM holes gas multiplication takes place (sections 3.2 and 3.3). Successively, part of the multiplied electrons drift in the transfer gap electric field towards GEM_2 where they are again multiplied. During this process, photons are emitted by the electron excited Ar/CF_4 molecules (section 3.4, 3.6) when they decay to the ground state. Photons are detected by means of a mirror-lens-CCD camera system (represented in Figure 3.1). The measured 2D light intensity distribution is proportional to the 2D distribution of the energy deposited in the drift gap.

Figure 3.3 On the left, the entrance window side of the scintillating GEM detector. On the right, exit window side of the chamber. Through the glass window, GEM_2 can be seen.

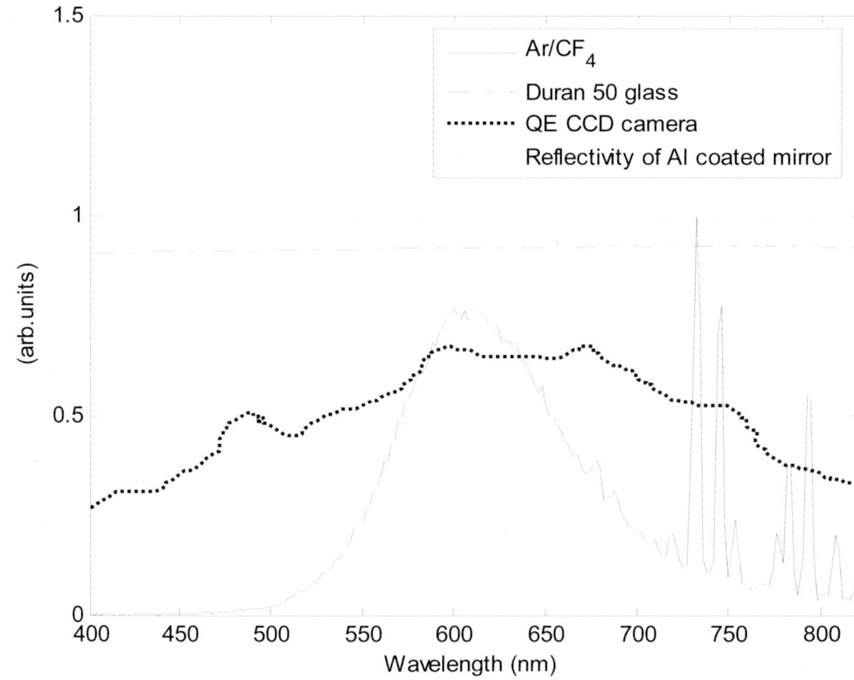

Figure 3.4 Ar/CF_4 92/08 emission spectrum; Duran 50 glass transmission; CCD camera quantum efficiency (QE) and aluminium coated mirror reflectivity. The Ar/CF_4 emission spectrum has been normalized to the highest spectrum value.

3.2 Gas amplification in a gas proportional counter

As already mentioned in Chapter 2, the passage of an ionizing particle through a gas creates ion pairs. Under the influence of an applied electric field, electrons begin to drift through the gas, while positive ions move in the opposite direction at much lower speed [4]. The drifting positive ions and free electrons diffuse by multiple collisions in the gas volume following a Gaussian law. During the migration of these charges, electrons might be captured by positive ions (*recombination*). Another process of electron capture occurs when a free electron meets a natural electronegative gas molecule, with the consequent formation of a negative ion (*electron attachment*). Charge transfer collisions can occur when a positive ion encounters a neutral gas molecule. In such a collision, an electron is transferred from the neutral molecule to the ion. An extensive discussion about all these processes can be found in [4].

The drifting electrons are accelerated by the electric field and they induce gas multiplication. The multiplication process is based on secondary ionization, or in other words on the ion-pairs created during the collision of the primary electrons with the neutral gas molecules, when the former have an energy greater than the ionization energy of the gas. Gas multiplication takes the form of a cascade, in which each electron created by ionization can potentially create more free electrons by the same process.

If n is the number of electrons at a given position x, after a path dx the number of electrons will be

$$dn = \alpha n dx \tag{3.1}$$

where α is the Townsend coefficient and represents the mean number of secondary electrons produced by a free electron per centimetre of its path length. The Townsend coefficient depends on the filling gas, the gas pressure and the electric field strength.

Integrating equation 3.1, the experimental growth of the electron avalanche is expressed by:

$$G \equiv \frac{n}{n_0} = e^{\int_{x_0}^{x_f} \alpha(x)dx} \tag{3.2}$$

where G is the gas gain, n is the number of electrons at a distance x_f, n_0 is the initial number of electrons at the initial distance x_0.
For a parallel plate geometry (constant electric field),

$$G \equiv \frac{n}{n_0} = e^{\alpha x} \tag{3.3}$$

There is a limit for the multiplication process given by the probability increase for secondary processes, such as avalanches induced by photon emission, and space-charge deformation of the electric field. These secondary processes can eventually lead to spark breakdown. A phenomenological limit for multiplication before spark breakdown is given by the Raether condition, $\alpha x \sim 20$, or $G \sim 10^8$. A more detailed discussion about the gas multiplication process can be found in for example [4].

3.3 Gas Electron Multiplier

The gas electron multiplier (GEM) [2], represented in Figure 3.5, consists of two metal (Cu) layers separated by a 50 μm-thick insulator (Kapton), chemically perforated[w] with a regular matrix of open channels, holes.

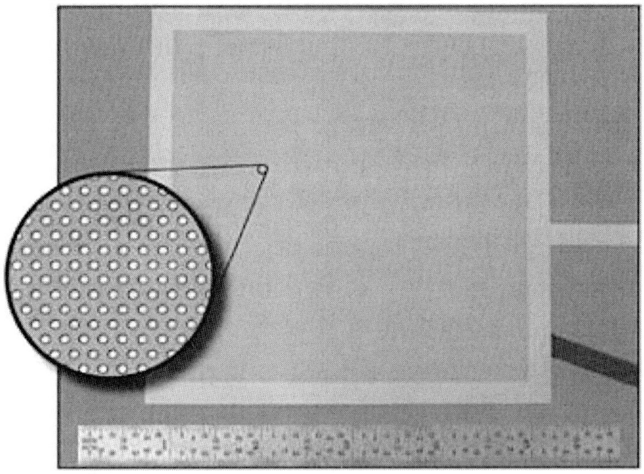

Figure 3.5 A raw GEM foil: the active area is 100x100 mm^2, with a 4 mm wide metal strip around edges and two contact leads on opposite sides; the Kapton foil outer dimensions are about 150x150 mm^2. With permission from the authors [1].

[w] A photolithography and acid etching process makes 30-50 μm diameter holes through both copper layers; a second etching process extends these holes all the way through the kapton.

The holes, arranged in the form of a hexagonal matrix, may be of any cross section, but are typically double-conical (Figure 3.6) with a diameter of between 25 μm and 150 μm, and a pitch of between 50 μm and 200 μm [5]. The GEM holes can be made very regular and dimensionally stable.

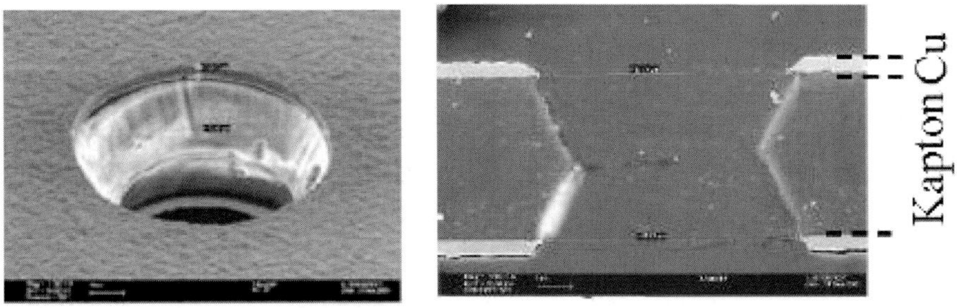

Figure 3.6 Detail of a double conical GEM hole: on the left, top view; on the right, hole cross section. With permission from the authors [1].

Application of a suitable potential difference between the two Cu sides of the GEM generates electric field lines similar to the ones represented in Figure 3.7.

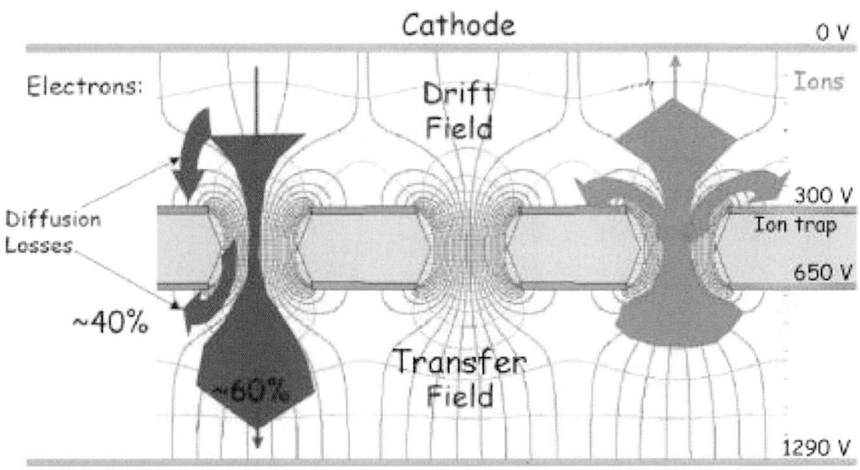

Figure 3.7 Working principle of a GEM detector. Modified from [6].

Placed in a chamber, containing a proportional counting gas, each hole will act as an independent proportional counter in which the avalanche is confined within a small space of a dimension of a few tens of microns. If an appropriate electrode-structure is added, most of the electrons released by ionization in the upper gas layer (drift gap), drift into the holes (amplification region) and multiply in the high electric field (50-70 kV/cm). Most of the avalanche electrons are transferred into the gas gap below the GEM (transfer gap). Some electrons are lost because they are "trapped" by the GEM surface at the highest voltage. The positive ions drift in the opposite direction of the electrons, towards the cathode. Some of these ions are collected by the GEM surface at the lowest voltage. The drift and transfer field must be optimized to avoid respectively recombination effects in the drift gap, and make sure as much electrons as possible are transferred to the next GEM (or anode). The GEM foil acts as a charge amplifier, preserving the original charge distribution to a large extent [7]. Figure 3.8 shows the field-strength distribution along the axis of a hole for different diameters [8]. As can be seen, the electric field in the hole approaches the field of a parallel-plate gap with decreasing hole diameter.

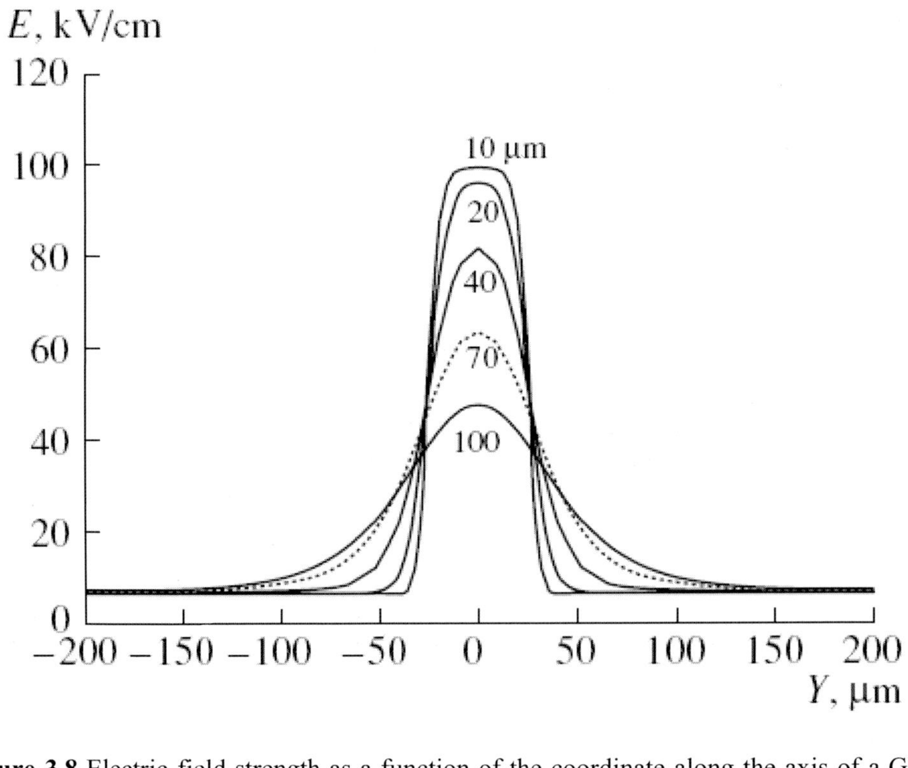

Figure 3.8 Electric-field strength as a function of the coordinate along the axis of a GEM hole for different diameters of the hole. $\Delta V_{GEM} = 500$ V, the dielectric (kapton) is 50μm thick, and the copper electrodes are 5 μm thick. With permission from the authors [18].

The maximum gain achievable with a GEM depends on the thickness of the polymeric support, the diameter of the holes, the gas mixture, and the applied voltages [8]. By cascading several GEMs, electrons are transferred from one GEM to the next, undergoing successive avalanches and yielding high gas gains [9]. The GEM transfer efficiency, defined as the efficiency with which an electron can be focused into one of the GEM apertures rather than being trapped by the metal entrance electrode [10], is a crucial issue, particularly in cascaded GEM detectors. The transfer efficiency depends on the given GEM geometry, the electric field above and across the GEM and on the gas type composition [10]. The transfer efficiency must be experimentally optimized in order to ensure optimal GEM operation [8].
A unique property of the GEM geometry is the complete separation of the amplification stage from the readout electrodes, which are usually very vulnerable to damage from sparking in the case of gas avalanche micro-pattern detectors [11]. Furthermore, due to its design, positive ion feed back into the drift region is reduced as compared to that of a wire chamber [12]. Another important advantage of GEMs is the high degree of granulation of amplification cells (50–100 holes/mm^2), which, in principle, allows imaging of highly resolved events during readout.
Another issue is the GEM detector gain variations with time: the presence of an insulator close to the multiplication channels introduces the possibility of dynamic gain shifts due to the deposition of charges (charge up) and the consequent modification of electric field [13, 14]. Possible solutions to this problem are: the addition of a small amount of water to the gas[x] [13], coating the insulator in the GEM holes with a very thin layer of slightly conducting material [15], use of cylindrical holes [16].
Among other applications [17,18], the gas electron multipliers can be used as an imaging device [19,20]: after multiplication in a proper gas mixture (section 3.5), the electrons still experience a high enough electric field , along part of their path outwards the GEM holes, to achieve scintillation light emission (section 3.4). This field region is basically confined to the exit part of the hole. The emitted light can be detected by an optical readout system [21]. First results on 2D dose imaging with a GEM detector are described in [22, 23].

[x] This solution can not be adopted when using a scintillating gas mixture such Ar/CF_4 because the light emission is influenced, namely partially inhibited, by the presence of humidity or water vapour.

3.4 Light production in gas proportional scintillation (GPS) counter

The gas proportional scintillation (GPS) counter is a hybrid detector which combines some of the properties of a proportional counter with those of a scintillation detector.

The GPS working principle is based on photons emitted by excited gas atoms or molecules. In a conventional gas scintillator, these excited gas molecules are created by direct interaction of incident radiation with the gas. The electronic excitation energy of these molecules is either dissipated non-radiatively by two-body collisions, or internal quenching processes or it is emitted as photons in the visible or ultraviolet (primary scintillation).

If an electric field is applied to the gas volume, the electrons from ion pairs, created along a particle track, drift as they do in a proportional counter. If the electric field is strong enough, inelastic collisions between these electrons and neutral gas molecules can elevate some of these molecules to excited states, which may de-excite through the emission of photons (secondary light). Figure 3.9 schematically represents the processes leading to electron impact light emission above discussed for a gas atom/molecule A.

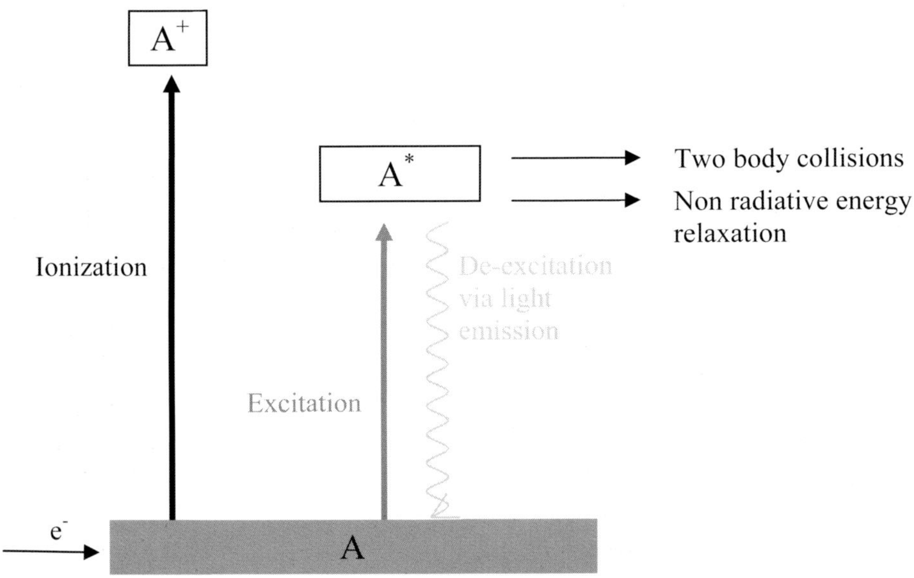

Figure 3.9 Schematic representation of the processes leading to electron impact light emission. A is a gas molecule or atom, A^+ is the ionized molecule/atom, A^* is the excited molecule/atom, e^- is an electron.

In GPS, the photon flux is proportional to the energy deposit of the charged particle being detected [24]. The secondary light magnitude increases with the increase of the electric field strength analogous to the one presented by the output charge of a proportional counter. The presence of molecular impurities in the gas inhibits the light emission. Therefore, the purity of gas is important for good secondary scintillation. The quenching of the light emitted by noble gases is stronger in the VUV region than for infrared emissions [24,25,26].

3.5 Gas mixture considerations

As mentioned in Chapter 1, the scintillating GEM detector is filled with a scintillating mixture of noble gas argon and CF_4. The filling gas in a proportional counter is usually a noble gas because such gases have a high ionization cross section at relatively low field strengths [4]. Among noble gases, argon is chosen because of its low cost. Moreover, noble gases do not react chemically with the detector components and are not electronegative. In fact, since gas multiplication is critically dependent on the migration of electrons, the filling gas must exhibit a low electron attachment coefficient [25].

Argon has been used since many years also in scintillating proportional counters. Most studies concern UV and visible light emission [27,28]. Remotely, the noble gases scintillation yield in the visible and/or near-infrared regions has also been studied [29] and it has been proven that scintillation in this wavelength light range is very useful when charge coupled device (CCD) cameras are used for readout [30]. Typically, the CCD camera quantum efficiency wavelength range extends from 400 nm up to 900 nm [31].

It is known that argon based detectors are not very stable and can not reach very high gains [4,25]. The addition of a small amount of polyatomic gases (quench gases) is helpful in reducing instabilities and proportionally loss caused by photon-induced effects. In [30,32,33], it is shown that the light emission is reduced by the addition of a quench gas, although the emission becomes foreseeable and stable in time.

Alternatively to the frequently used quench gas CO_2, CF_4 has been introduced, since its quenching rate for Ar light emission is lower than that measured in carbon-dioxide [34]. Besides, CF_4 is known to be a good photon emitter, with emissions both in the UV and visible spectral regions [35,36,37,38,39]. CF_4 has high drift velocity and low diffusion coefficients,

properties that make the gas suitable for imaging purposes [33, 40]. Several groups are currently using Ar/CF_4 mixtures in GEM detectors [41,42,43] and the scintillating properties of such mixture were studied at Coimbra University, Portugal [44,45].
In the work presented in this thesis, we used a mixture of Ar/CF_4 in a ratio of 92 volume percentage Ar and 8 volume percentage CF_4 (92/08). Experimentally it has been found that this Ar/CF_4 mixture gives the highest light output (Chapter 5).

3.6 Light scintillation in the scintillating GEM detector

In a GEM detector, filled with a scintillating gas mixture, the GEM holes are the place of scintillation light because of the higher electric field strength inside such channels [46]. In a double GEM detector, GEM_2 holes are expected to be the place of highest photon emission because at this detector stage there is the highest number of multiplied electrons. Moreover, the light emitted from GEM_1 holes is hidden by the presence of the second GEM.

3.6.1 Ar/CF_4 emission spectrum

The Ar/CF_4 92/08 emission spectrum recorded in the wavelength range 400-820 nm is represented in Figure 3.10. For a detailed description of the setup used for emission spectrum measurements see chapter 5.
The spectrum is characterized by a broad band with a maximum of intensity around 620 nm, and by several sharp lines between 720 and 820 nm. According to literature [45], the broad band in the visible region is attributed to the electronic excited states of CF_4 products, while the sharp lines[y] are attributed to the presence of Ar. The Ar/CF_4 emission spectrum has also a band in the UV region [45], but this spectral region is not of interest for this work.

[y] In atoms the only degree of freedom are translational and electronic. Therefore, in atomic gases photon emission occurs only due to electronic transitions. The electronic transitions result in characteristic series of lines in the emission spectrum.

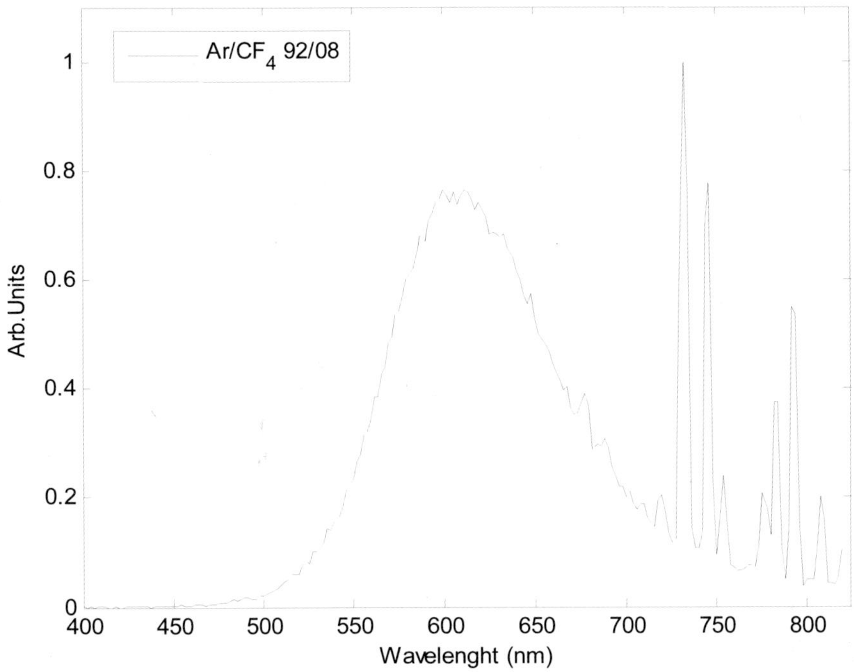

Figure 3.10 Emission spectrum of Ar/CF_4 92/08 normalized to the highest light intensity value.

The broad band results from the excitation of a Rydberg state[z] of the CF_4 molecule that dissociates into an emitting CF_3 fragment [47,48,49]. Details about the energy level diagram of the CF_4 molecule can be found in literature [50,51]. The near-infrared lines are Ar I atomic lines. They correspond to transitions between argon excited states ($3p^54p$) and ($3p^54s$) [45].

In the scheme represented in Figure 3.11, the processes leading to electron-impact photon emission in Ar/CF_4 mixture in the wavelength range 400-820 nm are summarized.

---

[z] A Rydberg state is characterized as a system where a single excited electron spends most of the time at so large distance from the ion core that the core can be represented almost as a point charge. It is this size, i.e., a radius much larger than core orbital, which qualitatively distinguishes a Rydberg orbital in a molecule. Since the Rydberg electron is nonbonding, the ion core will tend do dissociate as if the Rydberg electron were absent [54].

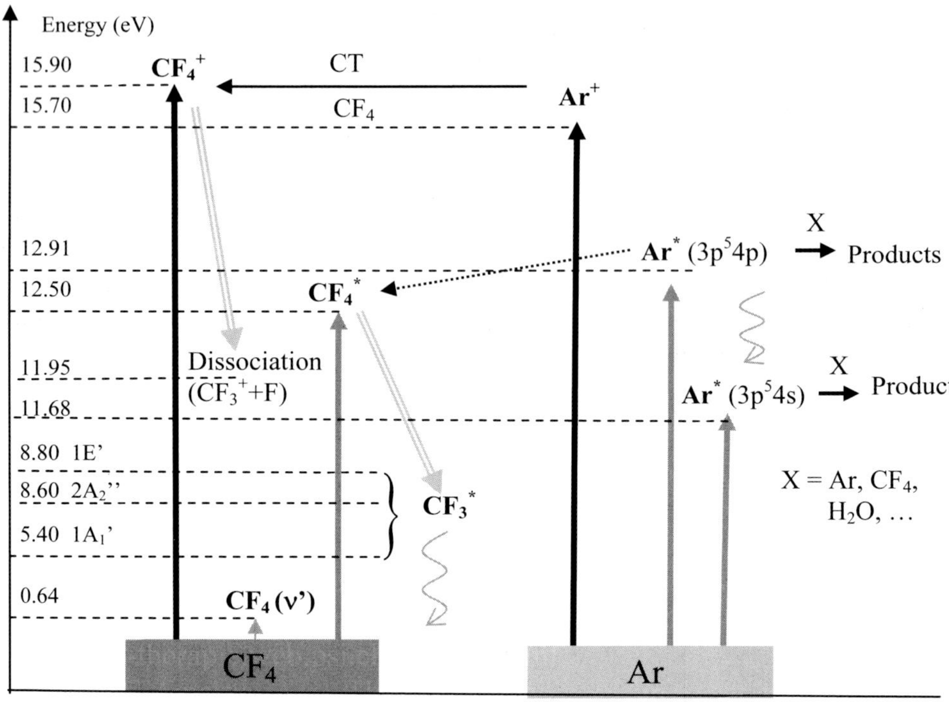

Figure 3.11 Schematic representation of the processes leading to photon emission due to electron impact on Ar/CF_4 in the wavelength range 400-820 nm.

The black thick solid vertical arrows indicate the ionization of CF_4 and Ar, while the grey thick solid vertical arrows show the excitation process.
The ionized CF_4, CF_4^+, dissociates promptly into daughter ions and the most probable products are CF_3^++F [49,52]. The emission of CF_3^+ is outside the studied wavelength range of 400 – 820 nm. According to literature [53], all electronic excited states of CF_4 and CF_4^+ seem to dissociate or predissociate with high probability. The excited CF_4, CF_4^*, has a very short life time [55], therefore only its dissociation products are seen, e.g. CF_3^*. The electronic excited states of CF_3^* are 1E', $2A_2$'', $2A_1$' [48]. The electronic transfer responsible of the visible emission is (1E', $2A_2$'' → $1A_1$'). $2A_2$'' and 1A1' are Rydberg states [56].
The excited Ar states decay emitting photons or undergo two-body collisions (indicated by horizontal black arrows pointing to the right) with for example Ar atoms, CF_4, and water molecules present in the mixture. Two-body collisions do not contribute to the light yield.

A charge transfer process (CT), indicated by the black thin horizontal arrow pointing to the left, occurs between the ionized argon and the CF_4 molecules, with consequent production of CF_4^+ [57] [aa].
An energy transfer is probably happening between excited argon and CF_4^* (indicated by the dot black horizontal arrow pointing left). Experiments show that the intensity of the visible molecular band does not vary much when the CF_4 concentration is increased in the mixture (see Chapter 5). This suggests that direct excitation of the CF_4 molecules by electron impact is not the main channel leading to the CF_3^* emission and an energy transfer mechanism between the excited states of argon and the dissociative electronic excited states of CF_4 may be present [45].

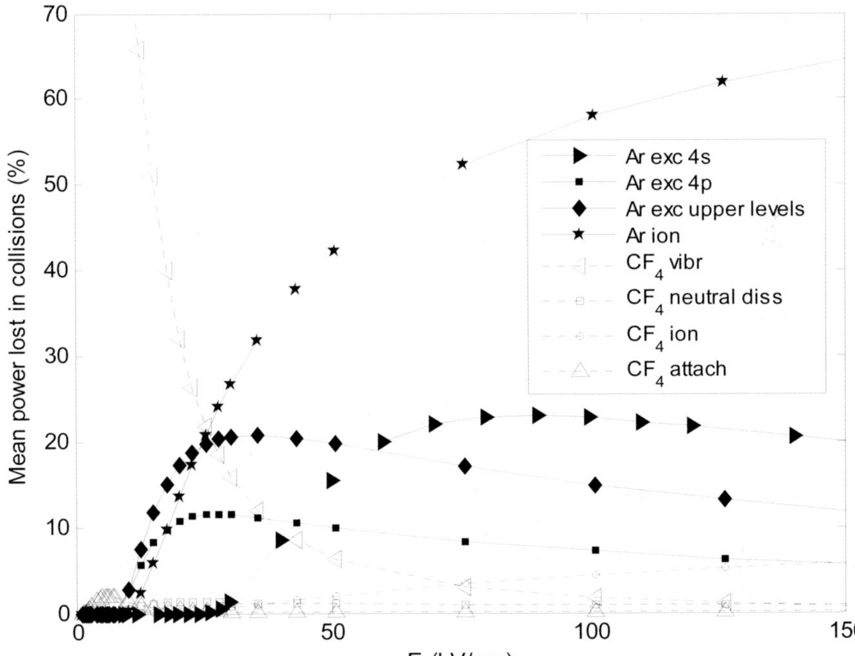

Figure 3.12 Probability (%) of a particular electron-impact process to occur in Ar/CF_4 92/08 as a function of the electric field. Data based on based on the numerical solution of the Boltzmann equation for a uniform electric field configuration, far from the detector walls and with the electrons in equilibrium with the electric field. Shown by courtesy of M. Fraga [58].

[aa] A direct electron transfer process from CF_4 to Ar^+ is not possible since the ionization potential of Ar is slightly lower than the CF_4 one, 15.7 eV against 15.9 eV. But if we assume that Ar^+ and CF_4 can form a complex ion $Ar^+ + CF_4 \rightarrow (Ar^+CF_4)^* \rightarrow Ar + F + CF_3^+ + 0.5$ eV the probability of the reaction with CF_4 could be higher.

For a quantitative idea of the occurrence of the electron-impact processes, the probability (%) that a particular electron-impact process occurs in Ar/CF_4 92/08 as a function of the electric field is represented in Figure 3.12. These calculations are based on the numerical solution of the Boltzmann equation for a uniform electric field configuration, far from the detector walls and with the electrons in equilibrium with the electric field.
In the electric field range typical of a GEM hole (50-70 kV/cm), among all electron impact processes, the ionization and excitation of argon are the most probable one. For both processes, the probability of occurring increases when the electric field is increased. Between 50 and 70 kV/cm, the probability of excitation increases more than the probability of ionization. Among the CF_4 electron-impact processes, ionization is most probable to happen in this electric field range, and its probability of occurring also increases when the electric field is increased.

3.7 Light signal readout

The photons produced during the gas multiplication process by the Ar/CF_4 excited gas molecules are detected by means of a low dark-current Apogee 1E camera coupled to a Tamron 171A zoom lens. The camera has a Kodak KAF-0401E (Apogee instruments) CCD with a quantum efficiency represented in Figure 3.4 that well matches the Ar/CF_4 emission spectrum. The camera is placed outside the beam to ensure low radiation background to the CCD. A 45° tilted aluminium coated mirror reflects the photons towards the camera (see Figure 3.13).
The reflectivity of the mirror is represented in Figure 3.4. The distance between the detector exit-window and the mirror is selected such to minimize reflections from the mirror back to the window. The light-path is enclosed in a light-tight plastic tube (Figure 3.13) that shields it from other light sources.
Some CCD camera specifications are listed in Table 3.1.
The camera digitizes to 14 bits, this means that it outputs 16348 levels (or gray level). The level is expressed in analog to digital unit (ADU). One ADU corresponds to 8.4 electrons generated by the collected photons (CCD camera gain). In the following chapters, the CCD signal per pixel is expressed in ADU.

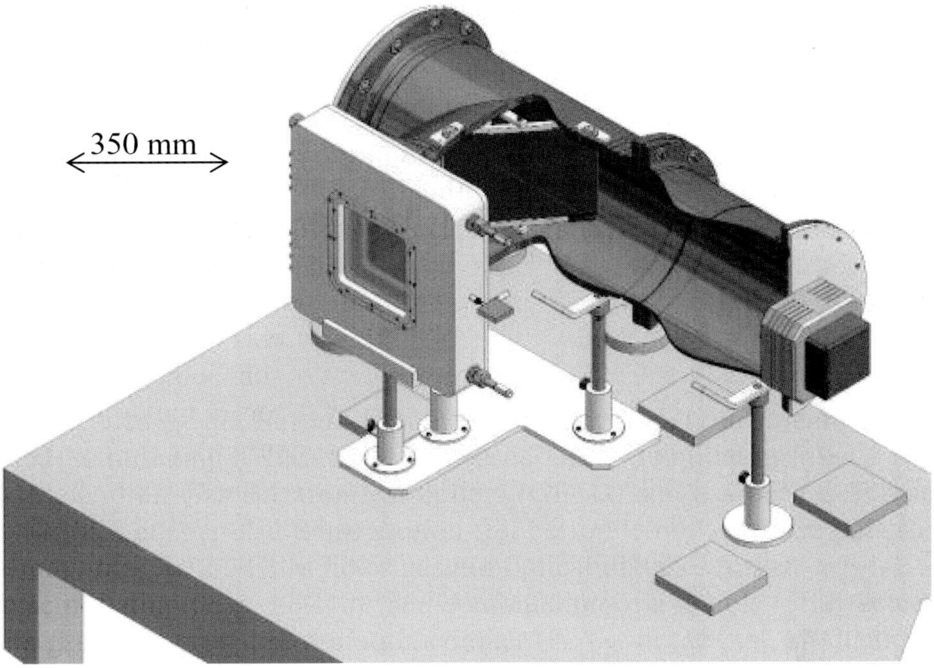

Figure 3.13 3D drawing, not to scale, of the scintillating GEM detector setup.

Array	768 × 512
Pixel size (μm)	9
Digital resolution	14-bit
Dark counts (e^-/pixel/s) at -20 °C	4.2[bb]
Gain (e^-/ADU)	8.4
Readout noise (e^-) at -20 °C	12.6[c]
Bias level (ADU/pixel)	1382

Table 3.1 Some CCD camera specifications.

[bb] The dark current and the readout noise have been experimentally determined and the values reported in Table 4.1 are in agreement with the respective values given by the CCD manufacturer. The dark current corresponds to the slope of the least linear square fit made on measured dark current values as a function of the exposure time. The readout noise is the standard deviation of several pictures taken for the minimum CCD camera exposure time (0.01 s).

The generation of the dark current is a thermal process wherein electrons use thermal energy to jump to an intermediate state, from which they are emitted into the conduction band. Electrons are excited independently of the light falling on the detector. These electrons are captured by the CCD potential wells and counted as signal. The most effective way to reduce the dark current is to cool the CCD, robbing electrons of the thermal energy required to reach an intermediate state. As a rule of thumb, the dark current halves with a decrease of 5-6 °C. In our case, the CCD camera is cooled down to – 20 °C during the measurements.

The signal-independent readout noise consists of two components. First, there is no perfect repeatability each time charge is dumped out of the CCD and digitized, or in other words the conversion from an analog signal to a digital number is not perfectly repeatable. Secondly, there is the injection of unwanted random signals by the sensor and electronics which end up getting digitized along with the pixel charge. The combination of these two random effects produces an uncertainty in the final output value for each pixel. In the output of the CCD picture, readout noise is added into every pixel every time the array is read out.

The bias level can also be thought as an offset. It can be determined measuring for the minimum CCD camera exposure time (0.01 s).

The CCD camera exposure time depends on the particular experiment. Usually it ranges between 30 and 100 s. In any case, it is always set slightly longer than the beam duration in order to guarantee the complete integration of the emitted light (see Figure 4.2).

The CCD camera is focused on GEM_2 by means of a 10 cm transparent foil, showing a 1cm-pitch grid, which is temporarily mounted at the GEM_2 location. The optical demagnification factor m of the whole set up is usually 0.043, given that 1 pixel (9μm × 9μm) on the CCD is equivalent to 210 μm × 210 μm at the grid position, and so at GEM_2 position.

3.8 The optical system

The photons emitted by the scintillating GEM detector N_{ph} are transported to the CCD camera by means of an optical system. Only a small fraction of these photons will reach the CCD sensor. The probability that a photon reaches the CCD camera and interacts with its pixels creating an electron can be expressed by:

$$k = k_1 \cdot k_2 \cdot k_3 \tag{3.4}$$

where k_1 is the probability that such a photon will be detected by the CCD camera, based on geometrical optics, k_2 is the probability that the photon will be transmitted by the optical elements, glass exit window, mirror, and lenses, and k_3 is the CCD camera quantum efficiency. The first term can be calculated using the following expression [59]:

$$k_1 = \frac{1}{16 \cdot F^2 \cdot (1 + 1/m)^2} \tag{3.5}$$

where F is the lens F-number (ratio of the focal length to the effective diameter), and m is the demagnification ratio (ratio of the linear object dimension to the linear picture dimension) [cc]. Typically during the experiments, the lens F-number was set to 3.8 and m was about 0.043, as specified in section 3.7. Therefore, the value of k_1 is about $7 \cdot 10^{-6}$.
k_2 is the product of three factors:

$$k_2 = k_{2,glass} \cdot k_{2,mirror} \cdot k_{2,lens} \tag{3.6}$$

where $k_{2,glass}$ represents the mean glass exit window transmission efficiency (Figure 3.4), (0.91 ± 0.01). $k_{2,mirror}$ is the mirror reflectivity (Figure 3.4), measured to be (0.99 ± 0.01). $k_{2,lens}$ is the lens transmittance; a value of 0.8 is typical [62,63]. Therefore, the value of k_2 is about 0.73.
For the third term of formula 3.4, a mean value of 0.5 has been evaluated by assuming a triangular shape of the quantum efficiency, represented in Figure 3.4, between 400 and 820 nm.
The product of all these factors leads to a k value of about $2 \cdot 10^{-6}$, the dominant contribution due to k_1.
The number of electrons per pixel $n_e(i)$ generated in the CCD chip by the photons emitted by the scintillating GEM detector, N_{ph} is

$$n_e(i) = k \cdot N_{ph} \tag{3.7}$$

[cc] According to [59], the optical refractive index of the scintillating medium n should also be taken into account. The refractive index of Ar/CF_4 is about 1 [60,61], therefore it can be neglected in equation 3.5.

In order to get the signal s(i) on a pixel in ADU , $n_e(i)$ must be multiplied by the inverse of the CCD camera gain (section 3.7), g.

$$s(i) = g \cdot n_e(i) = g \cdot k \cdot N_{ph} \sim 2 \cdot 10^{-7} \cdot N_{ph} \tag{3.8}$$

From equation 3.8, we see that is 1ADU is equivalent to about $0.5 \cdot 10^7$ photons. This equivalence has not been verified experimentally, therefore it is considered valid in first approximation only.

3.9 Electric signal readout

For a better understanding of the detector operation also the currents flowing onto the cathode and the GEMs surfaces are measured.
The cathode and all four GEM electrodes are connected to an individual channel of a CAEN high voltage power supply (SY127/A231). With respect to Figure 3.1, under normal operation conditions the cathode of the scintillating GEM detector is grounded; the electric fields in the drift gap, E_d, and in the transfer gap, E_t, are respectively 1 kV/cm and 1.5 kV/cm[dd] , unless mentioned otherwise, and the voltage across the GEMs is higher than 350 V (the upper limit of the applicable voltage across the GEM is related to the gas mixture and the type of radiation beam, see Chapter 5).
Since the currents flowing on the cathode and GEM surfaces are in a HV line, they can not be measured directly by means of data acquisition boards. In fact, the latter are incapable of measuring, at common mode, voltages larger than 10 Volt. Therefore, a High Tension Current Monitor (HTCM) was built in our electronics workshop for measuring the currents in high voltage lines, converting them into a proportional voltage to be measured by means of a PC-controlled National Instruments DAQ board. A simplified schematic of the HTCM is shown in Figure 3.14. With respect to Figure 3.14, HV_{out} is the line that goes to one of the GEMs surfaces and HV_{in} is the line coming from the power supply. The HTCM measures the currents flowing on the GEM surfaces by means of a resistor R_m in the high voltage line. The current value can be derived from the measured voltage drop across R_m. The voltage across R_m, sitting on top of a high voltage of about 1 kV, is measured by an isolation amplifier (AD210).

[dd] E_d=1 kV/cm, and E_t=1.5 kV/cm correspond to optimal drift and transfer field values for the scintillating GEM detector. E_d and E_t must be optimized before performing the experiments in order to respectively reduce the possibility of recombination and guarantee that a large number of electrons are focused on GEM_2 holes.

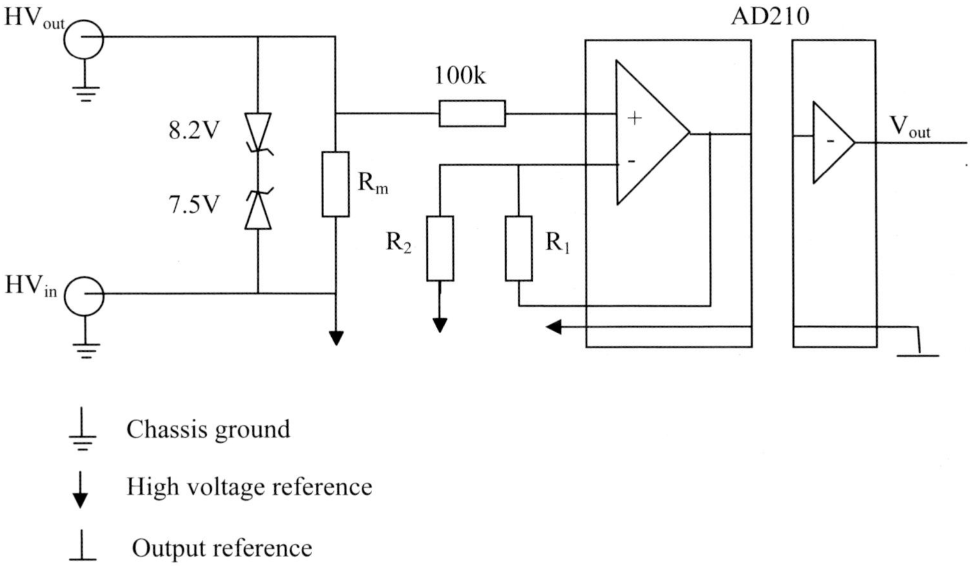

Figure 3.14 Basic scheme of the High Tension Current Monitor

This amplifier isolates the common mode high voltage component at its inputs and provides the amplified voltage difference across R_m at its output. Since the input circuit of the isolation amplifier is an opamp, the voltage gain can be set by means of the feedback network resistors R_1 and R_2. The input impedance of the isolation amplifier is very high (~ 5GΩ), therefore it hardly influences the voltage across R_m even if the latter is of the order of several MΩ. The intrinsic noise of the HTCM is 6 mV, independently on R_m value.

R_m values are chosen for each particular experiment in order to have a well detectable signal and at the same time a negligible voltage drop across them[ee].

A PC-controlled National Instruments DAQ board samples the voltage drops across R_m every 1 ms.

[ee] During the experiments described in chapter 5, 6, 8, R_m values were chosen in order to have a voltage drop in the each line supplying each GEM smaller than 0.25V. In fact, it was found that the voltage drop compensation on the detector outputs is less than 1% if the total voltage drop per GEM is smaller than 0.5V.

3.10 The gas flow system

The detector chamber is continuously flushed with a mixture of argon and CF_4 at 9 l/h and at 1 atm. The argon we normally use has a purity of 4.6 (99.996 %)[ff] while the CF_4 purity is 4.8 (99.998%). The two gases are mixed by means of a Brooks mixing station. The mixing station consists of two mass flow controllers (models 5850E, 5850S) and a read-out control unit (model 5878). After mixing the two gases in the specified ratio, the gas is flushed into the detector chamber by means of "polyflow" tubes. The gas leaving the chamber is lead to a pipe by means of which the gas is released outside the experimental area. Between the chamber outlet and the pipe a glass bubbler is inserted. The latter is filled with some oil. The bubbles that are created by the passage of gas give an indication that the gas is actually flowing.

3.11 References

[1] http://gdd.web.cern.ch/GDD/.
[2] F.Sauli, A new concept for electron amplification in gas detectors, *Nucl. Instrum. Meth.*, A 386,1997,531.
[3] http://www.national.com/mpf/LM/LM35.html.
[4] F.Sauli,*Principles of operation of multiwire proportional and drift chambers* CERN report , 1982.
[5] J.Benlloch et al., Development of the Gas Electron Multiplier (GEM), *IEEE Transactions on Nuclear Science*, 45, 1988, 234.
[6]http://www.ikf.physik.uni-frankfurt.de/IKF-HTML/highenergy/na49-lion/meetings/Cern-May-2005/Ropelewski/Ropelewski_collab_may05_gem_seminar-1.pdf.
[7]M.Titov, Micro-pattern gas detectors in Particle Detectors, http://pdg.lbl.gov/2007/reviews/pardetrpp.pdf, 2007.
[8] S. Bachmann et al., Charge amplification and transfer processes in the gas electron multiplier, *Nucl. Instrum. Meth.*, A 438, 1999, 376.
[9] S. Bachmann et al., Recent progress in GEM manufacturing and operation, *Nucl. Instrum. Meth.*, A 433, 1999, 464.
[10] C. Richter, A. Breskin, R. Chechik, D. Mörmann , G. Garty, A. Sharma, On the efficient electron transfer through GEM, *Nucl. Instrum.*

[ff] The purity is expressed as a percentage value, for example 99.996%. This is also abbreviated to 4.6, where the first digit refers to the number of nines and the last digit indicates the finest purity gradation.

Meth., *A* 478, 2002.
[11] F.D. Van De Berg, Gas-filled micro patterned radiation detectors, Thesis, Department of Radiation Technology, Interfaculty Reactor Institute, Delft University of Technology, (2000).
[12] F.Sauli, GEM readout of the time projection chamber. Internal report, CERN-EP-TA1, 1999.
[13] R.Bouclier et al., New Observations with the gas electron multiplier (GEM), *Nucl. Instrum. Meth.*, A 396, 1997, 50.
[14] P.J.Kramer, Charge up Effects of a GEM detector, Master Thesis, Department of Radiation Technology, Interfaculty Reactor Institute, Delft University of Technology, (2004), IRI/ST-ISO-2004015.
[15] S.Beirle et al., Carbon coated gas electron multiplers, *Nucl. Instrum. Meth.*, A 423, 1999, 297.
[16] J.Benlloch et al., Further developments and beam tests of the gas electron multiplier (GEM), *Nucl. Instrum. Meth.*, A 419, 1998, 410.
[17] F.Sauli, Development and applications of gas electron multiplier detectors, *Nucl. Instrum. Meth.*, A 505, 2003, 195.
[18] A. F. Buzulutskov, Radiation Detectors Based on Gas Electron Multipliers (Review), *Instruments and Experimental Techniques,* 50, 2007,287.
[19] F.A.F. Fraga, et al., Optical Readout of GEMs, *Nucl. Instrum. Meth.*, A 471 2001,125.
[20] F.A.F. Fraga, et al., CCD readout of GEM-based neutron detectors, *Nucl. Instrum. Meth.*, A 478, 2002, 357.
[21] F.A.F. Fraga et al., Imaging detectors based on the gas electron multiplier scintillation light, *Nuclear Science Symposium, 1999*. Conference Record 1999, IEEE Volume 2, 1999, 829.
[22] J.H.Timmer et al., A scintillating GEM for 2D-dosimetry in radiation therapy, *Nucl. Instrum. and Methods*, A 478, 2002, 98.
[23] S. Fetal et al., Dose imaging in radiotherapy with an Ar–CF_4 filled scintillating GEM, *Nucl. Instr. and Meth., A* 513, 2003, 42.
[24] A.J.P.L.Policarpo, Light Production and Gaseous Detectors, *Physica Scripta*,23, 1981, 539.
[25] G.F.Knoll, Radiation Detection and Measurement, 2nd Edition, John Wiley and Sons.
[26] E.Aprile et al, Noble Gas Detectors, Wiley-vch, 2006.
[27] P.E. Thiess and G.H. Miley, New Near-Infrared and Ultraviolet Gas-Proportional Scintillation Counters, *IEEE Transactions on Nuclear Science*, 21, 1974, 125.

[28] M. Susuki, et al., "The Emission Spectra of Ar, Kr or Xe+TEA", *Nucl. Instrum. Meth.,* A 254, 1987, 556.

[29] M.M. Fraga et al., Modelling of an IR scintillation counter, *Nucl. Instrum. Meth.*, A 442, 2000, 423.

[30] F.A.F Fraga, Quality control of GEM detectors using scintillation techniques, *Nucl. Instrum. Meth.*, A 442, 2000, 417.

[31] http://learn.hamamatsu.com/articles/quantumefficiency.html.

[32] M.M.Fraga, Study of Scintillation Light from Microstructure Based Detectors, *IEEE Transactions on Nuclear Science,* 47, 2000,993.

[33] A.Sharma, Properties of some gas mixtures used in tracking detectors, *ICFA* 16, 1998.

[34] A.Francis et al., Collisional deactivation of two photon excited Ar (4p; J = 0,2) states by H_2 and several hydrocarbon and fluorine containing molecules, *Eur. Phys. J. AP*, 4, 1998, 239.

[35] J. F. M. Aarts, "Ion and electron impact ionization of CF_4 studied via UV emission," *Chem. Phys. Lett.*, 114, 1985, 114.

[36] J. E. Hesser and K. Dressler, "Radiative lifetimes of ultraviolet emission systems excited in BF_3 , CF_4 and SiF_4 ," *J. Chem. Phys.*, 47, 1967, 344.

[37] U. Müller, T. Bubel, A. Sevilla, J. Dike, and K. Becker, "Further studies of the continuous UV emission produced by electron impact on CF4 ,"*Z. Phys. D*, 24, 1992, 131.

[38] L. C. Lee, J. C. Han, C. Ye, and M. Suto, "Fluorescence from photoexcitation of CF_3X (X = H, Cl, and Br) at 50–106 nm," *J. Chem. Phys.*, 92, 1990, 133.

[39] A. Pansky, A. Breskin, A. Buzulutskov, R. Chechik, V. Elkind, and J.Va'vra, The scintillation of CF_4 and its relevance to detection science, *Nucl. Instrum. Meth.*, A 354, 1995, 262.

[40] L. G. Christophorou, P. G. Datskos, and J. G. Carter, Gases of possible interest to SSC muon detectors, *Nucl. Instrum. Meth.*, A 309, 1991, 160.

[41] G.Bencivenni et al., Triple-GEM detector operation for high rate particle triggering, *Nuclear Physics B - Proceedings Supplements* ,125, 2003, 267.

[42] N. Abgrall , Characterization of a high resolution triple Gas Electron Multiplier (GEM) detector, *Nuclear Physics B - Proceedings Supplements*, 172, 2007, 234.

[43] M.Alfonsi, A triple-GEM detector for high-rate particle triggering *Nucl. Instrum. Meth.*, A 525, 2004, 17.

[44] M. Fraga et al., Pressure Dependence of Secondary NIR Scintillation in Ar and Ar/CF_4, *IEEE Transactions on Nuclear Science*, 48, 2001, 330.

[45] M.Fraga, The GEM scintillation in He-CF_4, Ar-CF_4, Ar-TEA and Xe-Tea mixtures, *Nucl. Instrum. Meth.*, A 504, 2003, 88.
[46] F.Fraga, Luminescence and imaging with gas electron multipliers, *Nucl. Instrum. Meth.*, A 513, 2003, 379.
[47] L.C.Lee at al., *J.Chem.Phys.,* 85, 1986, 6294.
[48] N.Washida, M.suto, Emission spectra of CF_3 radicals.IV. Excitation spectra, quantum yields, and potential energy surfaces of the CF_3 fluorescence, *J.Chem.Phys.*, 78, 1983, 1025.
[49] U.Muller et al., Photon/fragment-ion coincidence investigation of the continuous ultraviolet emission produced by fast ion impact on CF_4 and CHF_3, *J.Chem.Phys.*, 100, 1994, 5550.
[50] K.Furuya, et al., Fragment ion-photon coincidence investigation of carbon tetrafluoride by controlled electron impact, *J.Phys.B: At.Mol.Opt.Phys.*, 34, 2001, 1405.
[51] W.Zhang et al., Excitation and Ionization of Freon molecules.I.absolute oscillator strengths for the photoabsorption (12-740 eV) and the ionic photofragmentation (15-80 eV) of CF_4, *Chemical Physics*, 137, 1989, 391.
[52] K.Sthephan et al., Absolute partial and total electron impact ionization cross sections for CF_4 from threshold up to 180 eV, *J.Chem.Phys.*,83, 1985, 5712.
[53] L.G.Christophorou, J.K.Olthoff, Electron Interactions With Plasma Processing Gases: An Update for CF_4, CHF_3, $C2F_6$, and $C3F_8$, *J.Phys.Chem.Ref.Data,* 28, 1999, 197.
[54] H.F.Winter, M.Inokuti, Total dissociation cross section of CF4 and other fluoralkanes by electron impact, *Physical Review A*, 25, 1982, 1420.
[55] J.E.Hesser, K.Dressler, "Radiative Lifetimes of Ultraviolet Emission Systems Excited in BF_3, CF_4, and SiF_4^*, "*The Journal of Chemical Physics*, 47, 1967, 3443.
[56] I.Torres et al., Atomic fluorine emission cross sections observed in the 600-900 nm region following 0-500 eV electron impact on fluoromethanes (CF_4, CHF_3, CH_2F_2 and CH_3F), *J.Phys.B:At.Mol.Opt.Phys.* 32, 1999, 5437.
[57] M.Tsuij et al., Dissociative excitation of CF_4, CCl_4, and chlorofluoromethanes by collisions with argon and helium active species, *J.chem.Phys.*, 97, 1992, 245.
[58] M.Fraga, private communication.
[59] W. Swindell, The lens coupling efficiency in megavoltage imaging, *Med.Phys.*, 18, 1991, 1152.
[60] http://amelia.db.erau.edu/nasacds/200410Disc1/research/20040120957_2004116705.pdf;
[61] http://www.kayelaby.npl.co.uk/general_physics/2_5/2_5_7.html

[62] http://www.astrosurf.com/buil/us/spe2/hresol.htm#h1
[63] A. Karellas et al., Charge coupled device detector: performance considerations and potential for small field mammographic imaging applications, *Med.Phys.*, 19, 1992, 1015.

Chapter 4

Measuring procedures

In this chapter, the light and electric outputs of the scintillating GEM detector are defined. A classification is given of the background sources that can be found in a picture. The last part of the chapter deals with uncertainty analysis.

4.1 Irradiation setup

The characterization[gg] of the scintillating GEM detector (see Chapter 5) was performed in an x ray beam produced by an x ray generator with a copper anode. The x ray generator kV- and mA-values were set during the experiments in the range 12 – 24 kV and 0.75 – 15 mA respectively. As can be seen in Figure 4.1, the scintillating GEM detector was positioned, with the entrance window facing the x ray generator, downstream of a 3 cm diameter hole in a plate. During the experiments, the output of the x ray generator was constantly monitored by means of a PTW ionization chamber (0.69 cm^3) positioned in front of the beam coming from a different shutter with respect to the one used to irradiate the scintillating GEM detector. The integrated signal of this ionization chamber is named Ic.
The dosimetric properties of the scintillating GEM detector were verified in particle beams. In particular, the detector has been irradiated by a 360 MeV alpha beam and 150 MeV proton beam from the AGOR cyclotron at Kernfysisch Versneller Instituut, Groningen, the Netherlands. The alpha particle beam time period consisted of three days; the proton beam time period was eight days in 2005 and four days in 2006. The detector has been also tested in a clinical carbon ion beam at Gesellschaft für Schwerionenforschung mbH, Darmstadt, Germany. The effective beam time for the experiments in this beam was about five hours. A detailed

[gg] Characterization means the study of detector properties such to find best GEM hole size, the best Ar/CF_4 ratio, the proper E_d and E_t values, to study the emission spectrum etc. Some of the results of such detector characterization are discussed in chapter 5.

description of the proton, alpha and carbon ions irradiation setup can be found in respectively chapter 6, 7 and 8.

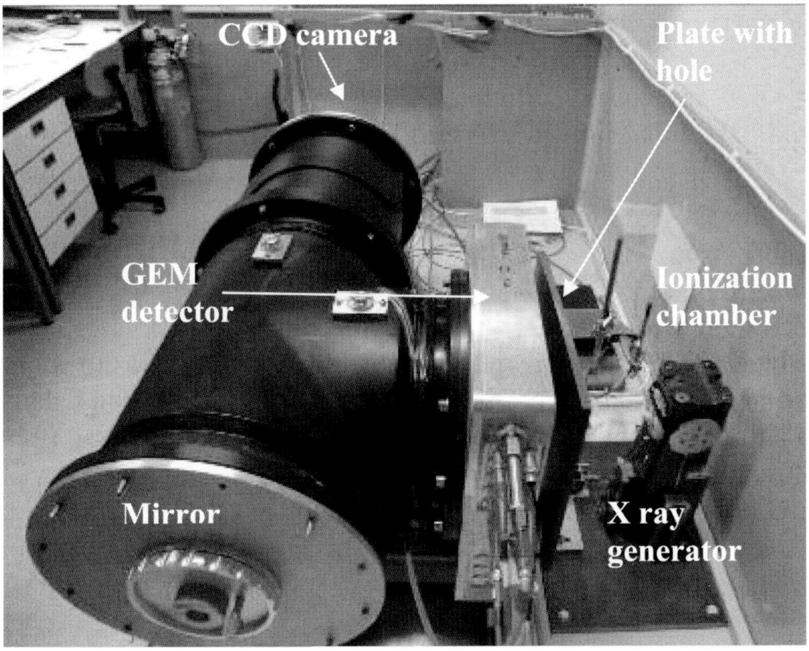

Figure 4.1 X ray irradiation setup.

For a single measurement, the beam is turned on to a particular beam intensity for a certain period of time. In that period, an amount of x rays or particles, corresponding to a predetermined amount of dose in water, is delivered to the scintillating GEM detector. The CCD camera shutter is open and the emitted light is integrated; the GEM detector electric signals and the beam monitor output, I_b, are sampled. Figure 4.2 shows an example of signals monitored during one measurement.

4.2 q_{out} definition

The electrons exiting the GEM_2 holes (Figure 3.1) are collected on the GEM_2 surface facing the exit window. The current measured on this surface is called I_{out}.
The offset, $< I_{offset} >$, is calculated from N_1 samples $I_{out}(t_i)$ before the beam start.

$$< I_{offset} >= \frac{\sum_{i=1}^{i=N_1} I_{out}(t_i)}{N_1} \tag{4.1}$$

The mean output current, $< I_{out} >$, is defined offline by averaging the sampled offset corrected output current values between instants t_s and t_f specified in Figure 4.2.

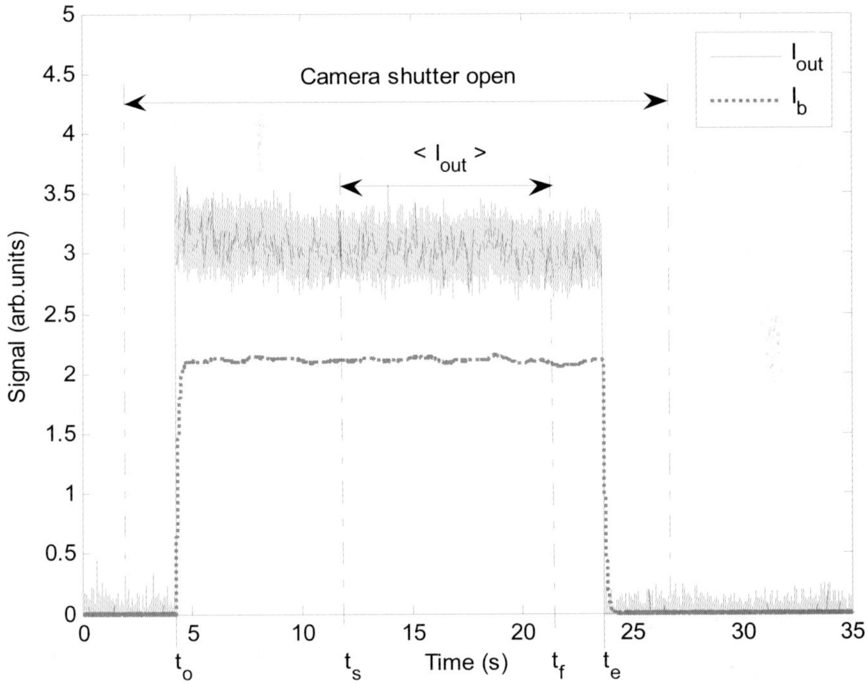

Figure 4.2 Graphical illustration of the signals recorded for each single measurement with the scintillating GEM detector as a function of time (x-axis). For visualization purposes of all the electric signals only I_{out} is plotted. Solid line: I_{out}; dashed line: beam monitor, I_b. CCD camera shutter open is indicated. t_o and t_e represent the beam-on and stop moments; t_s and t_f the instances in between which $< I_{out} >$ is evaluated. The camera exposure time is chosen such to have the shutter open for a time interval slightly longer than the beam duration to guarantee the complete integration of the emitted light.

$$< I_{out} >= \frac{\sum_{i=s}^{i=f} (I_{out}(t_i) - < I_{offset} >)}{n} \qquad t_s < t_i < t_f \tag{4.2}$$

where n is the number of samples measured between t_s and t_f.

The output charge, q_{out}, is the defined according to:

$$q_{out} = \left(\sum_{i=o}^{i=e} (I_{out}(t_i) - < I_{offset} >) \right) \cdot \Delta t \tag{4.3}$$

where t_o is the beam start time; t_e is the beam stop time; Δt is the sample time interval, 1 ms.

4.3 L_i definition

The CCD camera pictures are recorded and stored by means of a PC. The picture processing is made offline using Matlab routines.
The integrated light yield values, L_i, are obtained by integrating the picture pixels over a circular region of interest (ROI) after subtraction of a *background picture*.
Classification of background components that can be found on a picture is discussed in section 4.4. The chosen ROI is always bigger than the recorded beam spot on the picture. In Figure 4.3, an example of a measured picture is shown together with the light intensity profile along the one pixel-wide dashed white line. The continuous black line represents the region of interest on which the pixel values are integrated.

4.4 Picture background signals

In an picture *p* taken for a certain exposure time Δt_e and for a certain dose D in Gy, the following background sources can be identified:

- (**bk0**) The bias level or offset.
  It is a fixed number present in every picture that can be evaluated from an picture, bk0, measured for the minimum CCD camera exposure time (0.01 s).
- (**bk1**) The dark current.
  It depends on the CCD camera cooling and on the exposure time. It can be quantified taking a picture, bk1, for the same exposure time Δt_e, and the same cooling temperature of *p* but with the CCD camera shutter closed.
- (**bk2**) Direct interaction of scattered beam radiation with the CCD camera chip that results in large signals in isolated pixels (*spikes*). This background can be reduced by properly shielding the CCD camera, e.g. by means of lead bricks. It can be measured taking an picture, bk2, with

radiation beam on and CCD camera shutter closed for Δt_e seconds and D Gy.

- **(bk3)** Light emitted in the detector when the GEMs are off while E_d, E_t and the light gap electric field are on. This background component is due to the primary light emission (section 3.4) of the gas mixture and the glass exit window scintillation[hh]. It can be evaluated taking an picture, bk3, with radiation beam on, CCD camera shutter open for Δt_e s and D Gy, with GEMs off and E_d, E_t, light gap electric field on.

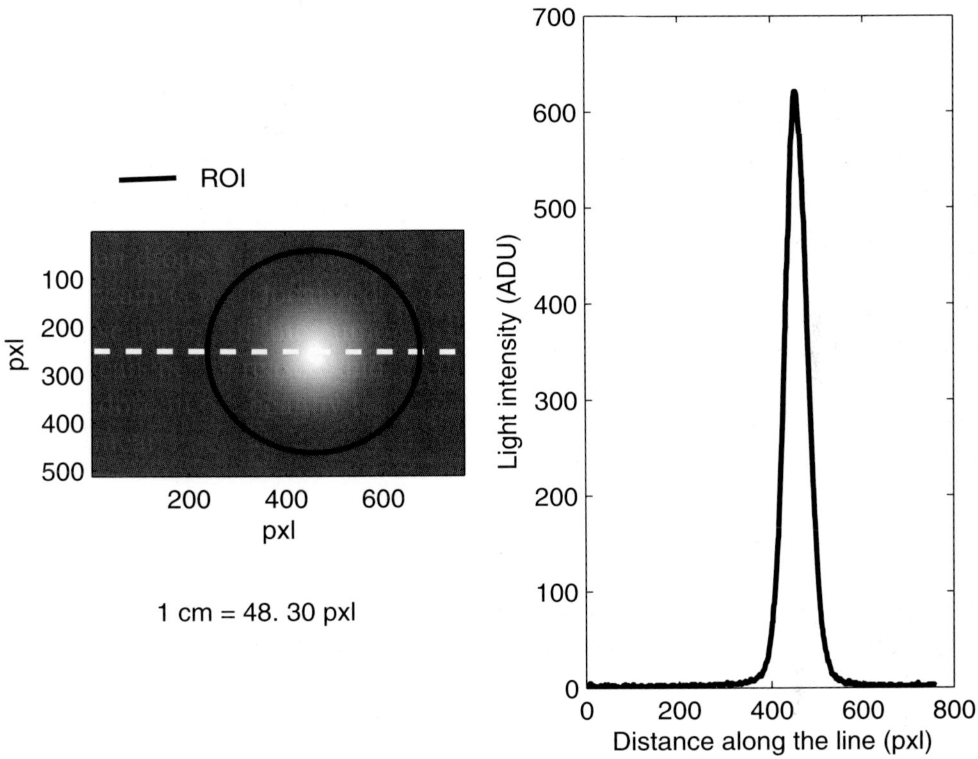

Figure 4.3 On the left, an example of an picture measured in a carbon ion beam is shown. The continuous black line represents the region of interest (ROI) in which the pixels are integrated. On the right, the light intensity profile along the one pixel-wide dashed white line is represented.

[hh] When E_d, E_t and the light gap electric field are on, a current I_{lg} is also measured on the surface of GEM_2 facing the exit window. It was found that the integrated I_{lg} signal, q_{lg} has the same sign of q_{out} and it increases linearly with the dose. q_{out} was not compensated for q_{lg} because the latter was found to be much smaller than q_{out}.

The bias level and the dark current are radiation beam independent background sources, while the remaining background components are radiation beam related.
The pixel contents of the picture *p* can be corrected for the background by subtraction. In principle, for a fixed exposure time Δt_e and dose D Gy, bk3>bk2>bk1>bk0; when bk3 is subtracted from *p*, also the other background components are automatically compensated.
In Table 4.1, typical light intensity values of a picture *p* measured in a proton beam and the main background sources are depicted.

	Light intensity (ADU/(pixel·Gy))
p	750.98
bk0 (bias level)	218.84
bk1-bk0 (dark current)	5.07
bk2-bk1 (*spikes*)	0.63
bk3-bk2 (gas/glass scintillation)	3.65
p-bk1	527.07
p-bk2	526.44
p-bk3	522.80

Table 4.1 Quantitative overview of the background sources listed above. The data are for proton measurements, 2005 run and for 6.4 Gy.

bk0 is about 30% of the *p* light intensity, while the dark current contribution, bk1-bk0, is about 1%. The gas and glass scintillation, bk3-bk2, is 0.5 % of the *p* light intensity. The background corrected signal is about 70% of the total light intensity of the picture *p*.

4.4.1 The background component due to light emitted when GEMs are off

When the voltage across the GEMs is zero but the electric fields in the drift, transfer and light gap are on, in presence of a radiation beam light is measured. The measured light is due to the Ar/CF_4 primary scintillation (section 3.4) and to the glass exit window scintillation. As already said, we call a picture taken under these conditions bk3.
We have found that bk3 increases linearly with the delivered dose, as shown in Figure 4.4.

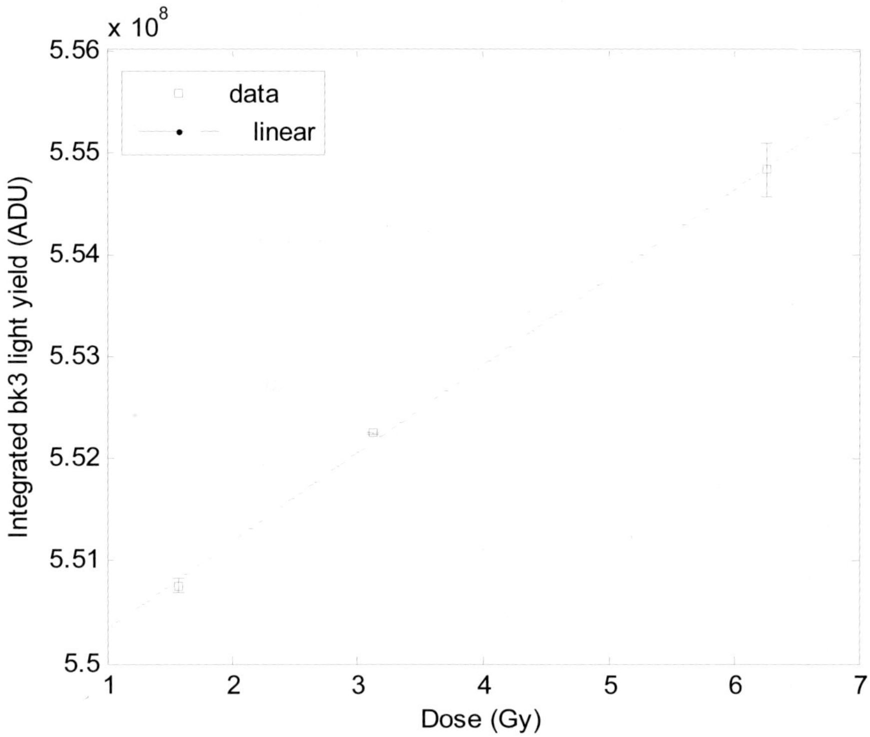

Figure 4.4 bk3 as a function of delivered dose

However, bk3 does not scale with the energy. In Figure 4.5, bk3 measured in a 150 MeV proton beam is shown as a function of the water depth[ii].
The data are normalized to the bk3 value obtained at the minimum water phantom thickness (0-wd). bk3 is almost constant till about 90 mm and then increases and its values fluctuate around a higher level. The bk3 increase at the Bragg peak depth (Bp-wd), ~ 108 mm, is not comparable with the expected peak to plateau ratio of 3.5 in the proton beam (see Chapter 6). This is due to the fact that the glass exit window scintillation is energy dependent analogous to the Lanex screen (see section 1.2.3) and its scintillation "quenches" the Bragg peak, smearing it out.

[ii] In order to measure bk3 as a function of the water depth, a picture was taken at the minimum water phantom thickness. Then, the thickness of the phantom was increased in steps up to a value larger than the proton range, and for each step a picture was taken.

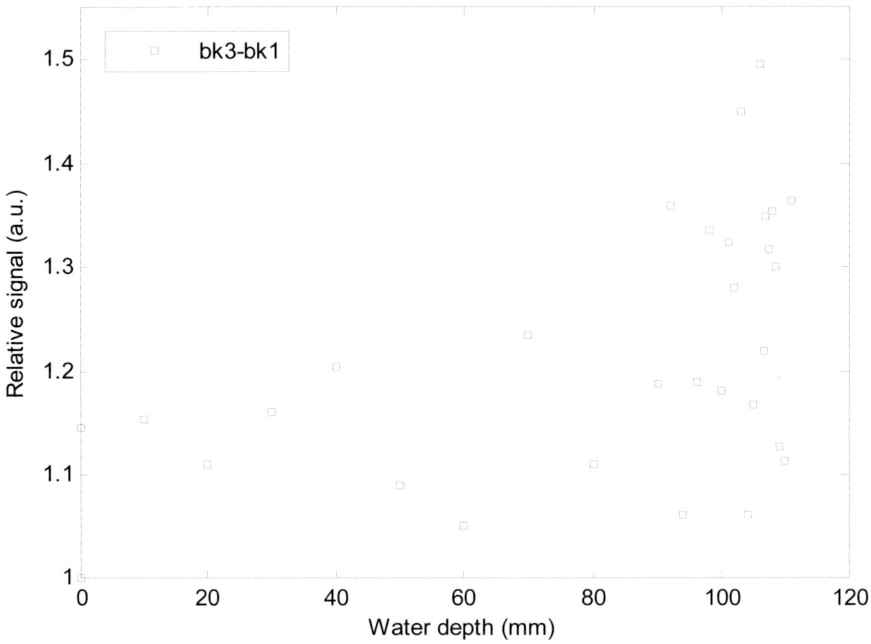

Figure 4.5 bk3 as a function of the water depth along a 150 MeV proton Bragg curve. The integration region of interest covers $1.6619 \cdot 10^5$ pixels. Error bars are not represented because they are smaller than the graph marker.

The fact that bk3 is energy dependent, and consequently it does not scale with the dose along a Bragg curve, represents a strong argument for measuring and subtracting this type of background from the pictures.

In Table 4.2, bk3 values measured in a carbon ion beam at 0-wd (bk3-0-wd) and at Bp-wd (bk3-Bp-wd) are reported together with the light intensity of the background corrected picture. The original picture p was measured at the Bragg peak depth of a ^{12}C beam. ((bk3-Bp-wd) – bk1) is slightly higher than ((bk3-0-wd)-bk1), as already observed in Figure 4.5 for proton beam data. This implies that p – (bk3-Bp-wd) < p – (bk3-0-wd).

	Light intensity (ADU/(pixel·Gy))
p	1693
(bk3-0-wd) – bk1	2.96
(bk3-Bp-wd) – bk1	3.19
(*p* – bk3-0-wd)	565.46
(*p* – bk3-Bp-wd)	565.19

Table 4.2 bk3 values measured at 0 mm water depth (0-wd) and Bragg peak depth (Bp-wd) are reported together with the light intensity of the background subtracted picture. The original picture *p* was measured at the Bragg peak depth of a ^{12}C beam. The data were measured in a ^{12}C beam and for ~ 1 Gy.

Figure 4.6 shows that bk3 is independent on the electric field in the light gap within the experimental errors, for a fixed delivered dose and a fixed water depth. This is expected because the glass scintillation is not electric field dependent and the primary Ar/CF_4 scintillation is constant for "low" electric field values, for example see [1].

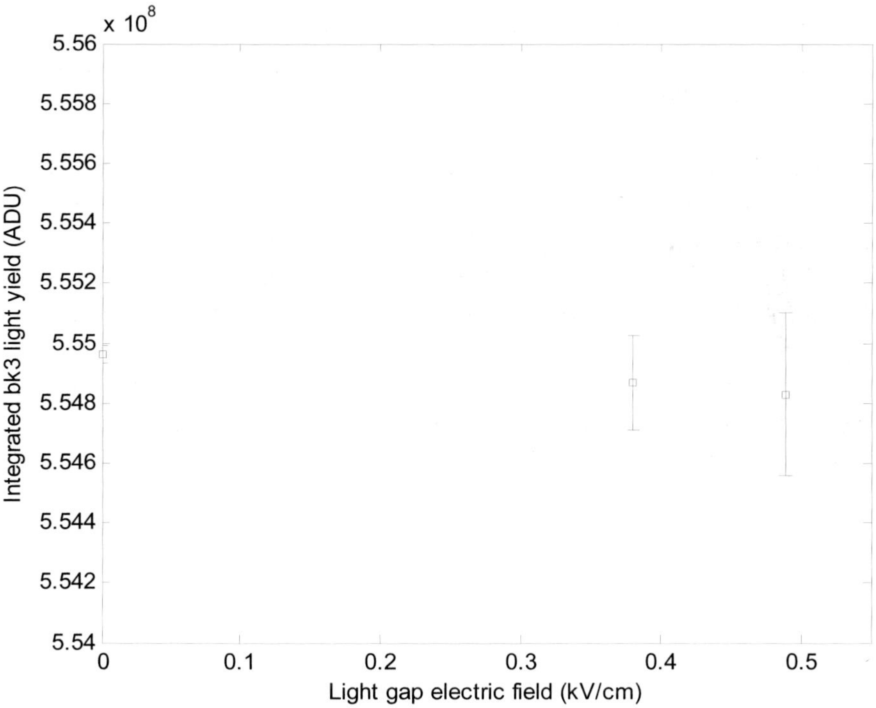

Figure 4.6 bk3 as a function of the electric field in the light gap.

4.4.2 Residual background

After correcting for bk3 background, a residual background is still present in the picture. The cause of this residual background is still not understood. It could be due to reflections. A quantitative idea of the residual background intensity can be obtained by integrating the pixel values in a region of an picture *p* where no signal is expected due to the radiation beam. For example, we consider the picture represented in Figure 4.7, from which bk3 has been subtracted, and we calculate the residual background in the picture corners. The picture was measured in a ^{12}C beam. The found residual background values are reported in Table 4.3, where corners are clockwise numbered starting from the upper left one. The intensity of this type of background is less than 1 % of the particular *p*-bk3 signal and the biggest difference among the different intensities is ~ 17 %. This means that it is not possible to compensate for the residual background by subtracting a single value.

Figure 4.7 Example of picture measured at the plateau of ^{12}C beam. The corners are clockwise numbered.

	ADU/(Gy·pixel)
p-bk3	513
Residual background - corner 1	1.88
Residual background - corner 2	1.59
Residual background - corner 3	1.64
Residual background - corner 4	1.91

Table 4.3 Residual background values calculated on the corner of the picture represented in

The importance of the different background sources with respect to the background subtracted light signal depends on the light intensity in the picture *p*. In other words, the higher the light signal the less important the background and so the less important which type of background is subtracted from the original picture *p*.
In Table 4.4, a background values overview is given for a typical picture taken in a proton beam (30 s exposure time, 10 Gy) and a typical picture taken in a ^{12}C beam (96 s exposure time, ~ 1 Gy).

	Light intensity ADU/(pixel·Gy)	
	Proton beam	^{12}C beam
p	381.09	1719
bk1-bk0	0.92	19.19
(bk3-0-wd)-bk1	0.83	2.56
p –bk1	243.0	516.12
p – (bk3-0-wd)	242.21	513.2
p –bk3 residual background	1.60	1.88

Table 4.4 Background overview of a picture taken in a proton measurement (2006 run) for 10 Gy, and one taken in ^{12}C measurements for ~1 Gy.

bk1-bk0 is about 0.2 % of the proton picture light intensity, while it is 1% for the carbon ion picture light intensity. (bk3-0-wd)-bk1 is about 0.2 % of *p* light intensity in both cases.
The residual background is respectively about 0.4% and 0.1% of *p* light intensity. In both cases, its magnitude is about the same. The light intensity of the background corrected picture taken in a ^{12}C beam is about 30 % of the *p* light intensity, while it is ~ 64% of the light intensity of a picture measured in a proton beam. The pictures in a proton beam were measured

for a detector gain that is twice than that in a carbon ion beam because in the latter beam the high voltage across the GEMs had to be reduced due to discharges. As a consequence the light signal recorded in the proton beam is about double with respect to the carbon ion light signal magnitude, as the light signal increases when the detector gain is increased (section 5.2.2).

4.5 Role of the median filter

The interaction of scattered radiation with the CCD camera results in large signals in isolated pixels (*spikes*). These *spikes* must be filtered out because they are not part of the "real" light signal and, if present in a picture, they modify its L_i value. A median filter (Matlab *medfilt2* routine [2]) is applied to a picture after background subtraction, in order to eliminate the spikes. A median[jj] filter is a non-linear filter that sets each output pixel to the median value of the neighbouring pixel intensities. We usually used a 3-by-3 median filter is used. In Figure 4.8, the light intensity profiles of a picture *p*, of the background corrected picture, *p*-bk3, and of *p*-bk3 with median filter applied are compared. In the picture *p* light intensity profile (continuous thin line), positive signals on single or few pixels are present. Some of these positive signals are *spikes* and some of them are due to "defected" pixels. When bk3 is subtracted from *p* (dashed line), the positive signals due the "defected" pixels disappear because these signals are present the same position also in the background picture. The positive signals due to the *spikes* remain. Moreover, the "negative signals" in the (*p*-bk3) profile are due to *spikes* in the background picture, not present in the original picture *p*. The median filter (continuous thick line) reduces the intensity of the *spikes*, making the light intensity profile smoother.

[jj] A median is the value that halves an ordered population. In our case, the ordered population are the pixel intensity values.

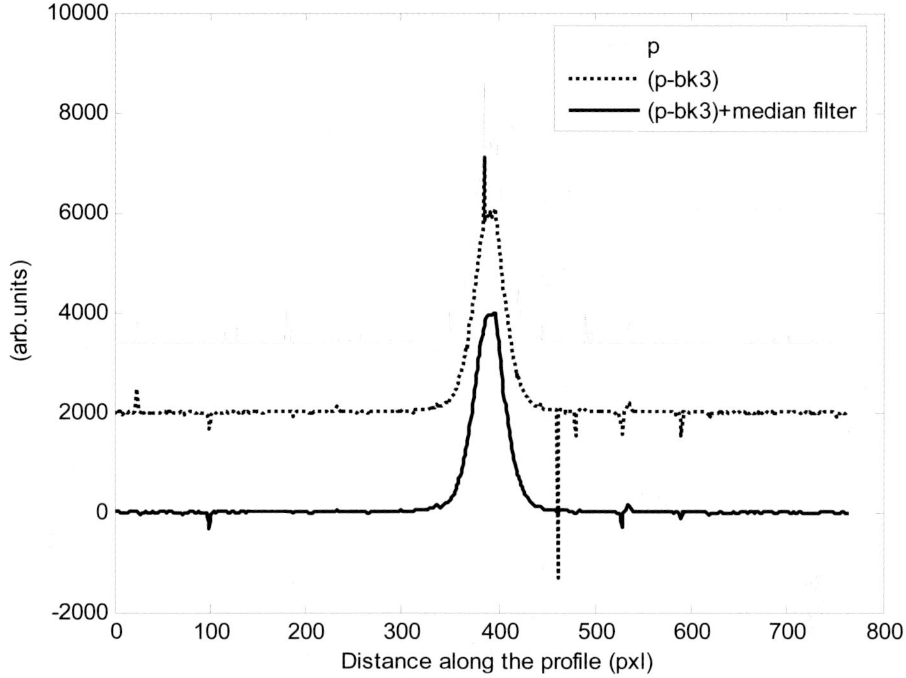

Figure 4.8 Comparison of light intensity profiles of a picture *p* (continuous thin line), the background corrected picture, *p*-bk3 (dashed line), and *p*-bk3 with the median filter applied (continuous thick line). The data are for proton measurements, 2006 run. For visualization purposes, a constant quantity has been added to the light intensity values of *p* and (*p*-bk3) profiles.

4.6 Error analysis

4.6.1 q_{out} uncertainty

The error on the offset (equation 4.1) is calculated according to the following formula:

$$\sigma_{<I_{offset}>} = \frac{1}{\sqrt{N_1}} \sigma_{I_i(off)} \tag{4.4}$$

where $\sigma_{I_i(off)}$ is the standard deviation of the sampled values on which the offset is evaluated.

The statistical error on q_{out} (equation 4.3) is evaluated by means of the error propagation formula:

$$\sigma_{q_{out}} = \Delta t \cdot \sqrt{N_2 \cdot \sigma^2_{I_i(on)} + N_2^2 \sigma^2_{<I_{offset}>}} \qquad (4.5)$$

where $\sigma_{I_i(on)}$ is standard deviation of the sampled values when the beam is on.

A distinction is made between $\sigma_{I_i(off)}$ and $\sigma_{I_i(on)}$ because it was found that the noise of the sampled values is larger when the beam is on. Probably this is due to interference picked up, e.g. from the accelerator system.

The experimental error of q_{out} is evaluated considering n measurements, performed in succession under the same measurement conditions, and evaluating the maximum q_{out} variation among them. The short term reproducibility error is defined as the biggest variation found in q_{out} values for data collected during a certain period of time, e.g. days apart.

In Table 4.5, a numerical comparison between the statistical error, the experimental error and the short term reproducibly error of q_{out} is reported. The statistical error is the maximum value of the uncertainty obtained using formula 4.5 for the q_{out} values of ten measurements performed in succession. The experimental error is evaluated on the same ten measurements performed in succession; while the short term reproducibility error for measurements performed at different moments during one week. The comparison is made in terms of relative error.

	Relative error (%)
Statistical error	0.3
Experimental error	1.5
Short term reproducibility error	4

Table 4.5 Comparison among the relative statistical error, the relative experimental error and the relative short term reproducibly error of q_{out}. Data are for proton measurements, 2005 run.

The experimental and short term reproducibility errors are bigger than the statistical error. This is due to the fact that the experimental and short term reproducibility errors include "extra" uncertainty sources. Data are normalized to the beam monitor signal, therefore beam fluctuations are excluded here as error source.

The short term reproducibility error is bigger than the experimental error. Ambient pressure variation can not explain this difference because the data on which the short term reproducibility is evaluated were corrected for that (see Chapter 5).
In the following, the experimental error is associated to q_{out} for measurements performed in succession or within a very short period of time, while the short term reproducibility error is associated to q_{out} measurements performed in quite different moments, e.g. days apart. Error bars are only shown when graph markers are smaller than the experimental uncertainties.

4.6.2 L_i error analysis

4.6.2.1 CCD camera noise sources

As already mentioned in section 3.7, the signal s(i) on each pixel from the CCD chip is read out in the adc unit ADU. s(i) is proportional to the charge on each pixel of the chip expressed in number of electrons, $n_e(i)$. The proportionality constant is the inverse of the CCD gain, g (Table 3.1).

$$s(i) = g \cdot n_e(i) \tag{4.6}$$

The noise associated to s(i) is read out with the same factor g.

$$\sigma_{s(i)} = g \cdot \sigma_{n_e(i)} \tag{4.7}$$

The number of electrons $n_e(i)$ is formed by the charge generated by the photons falling onto the chip $n_e^{ph}(i)$, the electrons from the dark current $n_e^{dk}(i)$, and the bias level or offset charge $n_e^{offset}(i)$ (see section 3.7).

$$n_e(i) = n_e^{ph}(i) + n_e^{dk}(i) + n_e^{offset}(i) \tag{4.8}$$

The noise $\sigma_{n_e(i)}$ on a pixel associated with the acquisition of a picture by the CCD consists of [3,4,5]:

- The noise due to the random arrival of photons at any detector. It is the product of the CCD camera quantum efficiency k_3 (defined in section 3.8) and the square root of the number of collected photons (since the time between photon arrivals is governed by Poisson statistics).

$$\sigma_{n_e^{ph}(i)} = k_3 \cdot \sqrt{n^{ph}(i)} \tag{4.9}$$

- The dark current noise, $n_e^{dk}(i)$. Although the dark current signal (section 3.7) can be corrected for, the noise associated with this signal can not. The dark current noise is equal to the square root of the dark current signal,

$$\sigma_{n_e^{dk}(i)} = \sqrt{n_e^{dk}(i)} \tag{4.10}$$

- The readout noise, $\sigma_{n_e^{offset}(i)}$, described in section 3.7.

Therefore, according to the error propagation formula

$$\sigma_{n_e(i)} = \sqrt{\left(\sigma_{n_e^{ph}(i)}\right)^2 + \left(\sigma_{n_e^{dk}(i)}\right)^2 + \left(\sigma_{n_e^{offset}(i)}\right)^2} =$$

$$= \sqrt{k_3^2 \cdot n^{ph}(i) + n_e^{dk}(i) + \left(\sigma_{n_e^{offset}(i)}\right)^2} \tag{4.11}$$

In our case, a background picture is always subtracted from the original picture. $s_p(i)$ is the signal on a pixel in ADU of a picture *p*, $s_b(i)$ is the signal on a pixel in ADU of a background picture. $\tilde{s}_p(i)$ is the signal on a pixel in ADU of the background subtracted picture p.

$$s_p(i) = g \cdot n_{e,p}(i) = g \cdot (n_{e,p}^{ph}(i) + n_{e,p}^{dk}(i) + n_{e,p}^{offset}(i)) \tag{4.12}$$

$$s_b(i) = g \cdot n_{e,b}(i) = g \cdot (n_{e,b}^{ph}(i) + n_{e,b}^{dk}(i) + n_{e,b}^{offset}(i)) \tag{4.13}$$

$n_{e,p}(i)$ and $n_{e,b}(i)$ are the number of electrons respectively in the picture *p* and in the background picture.

Since the background picture is always measured for the same exposure time and temperature of the picture p, $n_{e,p}^{dk}(i) = n_{e,b}^{dk}(i) = n_e^{dk}(i)$ and $n_{e,p}^{offset}(i) = n_{e,b}^{offset}(i) = n_e^{offset}(i)$. So,

$$\tilde{s}_p(i) = s_p(i) - s_b(i) = g \cdot (n_{e,p}^{ph}(i) - n_{e,b}^{ph}(i)) \tag{4.14}$$

The noises associated to the picture p, $\sigma_{s_p(i)}$, and the background picture, $\sigma_{s_b(i)}$, are defined as:

$$\begin{aligned} \sigma_{s_p(i)} &= g \cdot \sqrt{\left(\sigma_{n_{e,p}^{ph}(i)}\right)^2 + \left(\sigma_{n_{e,p}^{dk}(i)}\right)^2 + \left(\sigma_{n_{e,p}^{offset}(i)}\right)^2} = \\ &= g \cdot \sqrt{k_3^2 \cdot n_p^{ph}(i) + n_{e,p}^{dk}(i) + \left(\sigma_{n_{e,p}^{offset}(i)}\right)^2} \end{aligned} \tag{4.15}$$

$$\begin{aligned} \sigma_{s_b(i)} &= g \cdot \sqrt{\left(\sigma_{n_{e,b}^{ph}(i)}\right)^2 + \left(\sigma_{n_{e,b}^{dk}(i)}\right)^2 + \left(\sigma_{n_{e,b}^{offset}(i)}\right)^2} = \\ &= g \cdot \sqrt{k_3^2 \cdot n_b^{ph}(i) + n_{e,b}^{dk}(i) + \left(\sigma_{n_{e,b}^{offset}(i)}\right)^2} \end{aligned} \tag{4.16}$$

According to formulas 4.15 and 4.16, the noise of the background subtracted picture $\sigma_{\tilde{s}_p(i)}$ is

$$\begin{aligned} \sigma_{\tilde{s}_p(i)} &= \sqrt{\sigma^2_{s_p(i)} + \sigma^2_{s_b(i)}} = \\ &= g \cdot \sqrt{k_3^2 \cdot n_p^{ph}(i) + k_3^2 \cdot n_b^{ph}(i) + 2 \cdot n_e^{dk}(i) + 2 \cdot \left(\sigma_{n_e^{offset}(i)}\right)^2} \end{aligned} \tag{4.17}$$

The $\tilde{s}_p(i)$ is always integrated over a region of interest, as described in section 4.3.

For example, if we chose a region of interested with N pixels, the integrated light yield L_i is

$$L_i = \sum_1^N \tilde{s}_p(i) = \sum_1^N g \cdot (n_{e,p}^{ph}(i) - n_{e,b}^{ph}(i)) \tag{4.18}$$

The error associated to L_i, σ_{L_i} is

$$\sigma_{L_i} = \sqrt{\sum_1^N \sigma^2_{\bar{s}_p(i)}} =$$

$$= \sqrt{\sum_1^N g^2 \cdot (k_3^2 \cdot n_p^{ph}(i) + k_3^2 \cdot n_b^{ph}(i) + 2 \cdot n_e^{dk}(i) + 2 \cdot \left(\sigma_{n_e^{offset}(i)}\right)^2)} \qquad (4.19)$$

In equation 4.19, $\sum_1^N g^2 \cdot k_3^2 \cdot n_b^{ph}$ is zero if the background is measured with shutter closed, i.e. bk1.

4.6.2.2 Uncertainty in L_i

The experimental error of L_i, likewise the q_{out} experimental error, is calculated considering the maximum variation of L_i values among n measurements performed in succession. The short term reproducibility error is defined as the biggest variation found in L_i values for data collected during a certain period of time, e.g. days apart.
In Table 4.6, a numerical comparison between the statistical error, the experimental error and the short term reproducibly error of L_i is reported. The statistical error is the maximum value of the uncertainty obtained using formula 4.19 for L_i values of ten measurements performed in succession. The experimental error is evaluated for the same ten measurements; while the short term reproducibility error for measurements performed at different moments during one week. The comparison is made in terms of relative error.

	Relative error (%)
Statistical error	0.01
Experimental error	1.8
Short term reproducibility error	5

Table 4.6 Comparison among the relative statistical error, the relative experimental error and the relative short term reproducibly error of L_i. Data are for proton measurements, 2005 run.

L_i statistical error is quite small compared to the experimental error and the short term reproducibility error, as also found for q_{out} errors. The data used to calculate the experimental and shot term reproducibility error were

compensated for beam fluctuations and ambient pressure variations. The experimental and short term reproducibility relative errors of L_i and q_{out} (Table 4.5) are similar. This means that the gas multiplication process, responsible of the creation of both detector outputs, is introducing some sources of uncertainty. Why for both detector outputs, the short term reproducibility is bigger than the experimental error is still an open question. As for $\sigma_{q_{out}}$, in the following the experimental error is associated to L_i values for measurements performed within a short period of time, while the short term reproducibility error is associated to L_i values for measurements made in quite different time period, e.g. days apart.
Error bars are only shown when graph markers are smaller than the experimental uncertainties.

4.6.2.3 Considerations on the light signal magnitude

The accuracy in detecting a certain dose value depends on the magnitude of the light signal. We consider a typical proton beam background corrected light intensity of 242.21 ADU/(pixel·Gy) (Table 4.4), 30 s exposure time and a region of interest with $N = 2 \cdot 10^4$ pixels. According to equation (4.19), σ_{L_i} is about 0.2 % when measuring 5 cGy (5 cGy is the lowest experimentally measured dose, see section 6.3.3). The main component of σ_{L_i} is in this case due to the dark current noise. When detecting 1mGy under the same conditions, the L_i statistical uncertainty is about 8 %, the dark current and the readout noise contributions being much higher than the noise associated with the random arrival of photons at the CCD. In order to decrease the L_i uncertainty, the light signal magnitude must be increased since the dark current and the readout noise contributions remain the same if the exposure time and the region of interest area are unchanged. If, for example, the light signal intensity is increased by a factor of 10, then the L_i error drops to 0.8 % when measuring 1mGy.
The light signal can be increased, simply decreasing the lens F-number. According to equation 3.5, if lenses with an F-number of 1 are chosen instead of the actual 3.8, the probability that a photon is detected increases by a factor of about 14.

4.7 References

[1] A.Pansky, the scintillation of CF_4 and its relevance to detection science, *Nucl. Instrum. Methods*, A 354, 1995, 262.
[2] Matlab MATLAB® - The Language of Technical Computing, Mathworks Inc.,Natick MA, version 7.0.0.19920 (R14).
[3] CCD Image Sensor Noise Sources, application note, w w w . k o d a k . c o m / g o / i m a g e r s.
[4] Reibel et al., CCD or CMOS camera noise characterization, *Eur. Phys. J.,* AP 21, 2003.
[5] Y. Sutcu et al., Improvements on sensor noise based source camera identification, , presented at International conference on multimedia & Expo, IEEE, July 2-5, 2007, Beijing, China.

Chapter 5

Scintillating GEM detector characterization in an x ray beam

In this chapter, the experiments performed to characterize the scintillating GEM detector under x ray irradiation are described. X rays were chosen because of its easy access and daily availability in the laboratory, unlike a particle beam. Anyhow, the results reported below are considered to be radiation-type independent.

5.1 Experimental setup

With respect to the scintillating GEM detector setup described in Chapter 3, it must be added that in order to measure the primary charge needed to calculate the detector gain (section 3.2), two extra aluminized Mylar planes were inserted in the detector chamber between the entrance window and the already existing cathode. The first plane is a cathode, C_1, and the second a anode, A. The distance between the entrance window and C_1 is 0.5 mm; the distance between C_1 and A, and the distance between A and the already existing cathode were set equal to the drift gap. The anode A was supplied with high voltage, while the other two cathodes were grounded. The primary charge, q_p, was calculated according to formula 4.3, where instead of I_{out}, the primary current I_p measured on the anode A, was used. It was verified that I_p was about twice the current flowing on the C_1 cathode. This is expected because the gap thickness between C_1 and A, and A and the original cathode is the same.
Moreover, for the pressure "calibration" curve experiment (section 5.2.4), a needle valve was inserted in series to the exhaust gas tube between the detector chamber and the glass bubbler. The needle valve was used to vary the pressure, in the range 990-1028 mbar, inside the detector chamber with respect to the atmospheric pressure.
For the emission spectrum measurements, the Ar/CF_4 emitted light was analyzed by a Macam monochromator (mod. MCG 910) equipped with a

1200 grooves/mm grating (8 nm/mm dispersion and about 3.5 nm resolution) and recorded by means of a Hamamatsu R943-02 photomultiplier tube covering the wavelength region 250 - 850 nm. The photomultiplier tube was supplied with -1700 V, cooled down to -20° C and operated in single counting mode. Emission spectra were collected in the range 400 – 820 nm in steps of 2 nm and 4 s integration time per step. The wavelength range was chosen taking into account the sensitivity of the CCD camera (because this is normally used as readout). The light path from the scintillating GEM detector to the monochoromator and the photomultiplier tube was properly shielded in order to avoid ambient light contamination.
The spectra reported in the following are corrected for the sensitivity of the setup, unless mentioned otherwise. The sensitivity curve was measured by means of a calibrated EPLAB 1000Watt Quartz Iodine Lamp. We did not place a cut-off filter at the entrance slit of the monochromator for eliminating second-order diffraction effects because negligible differences were found between two spectra measured with and without a long-pass colour glass filter with 435 nm cut-off wavelength.
L_i values and the integral values of the emission spectra have not been compared because the light collection angle of the lens system is different from the one of monochoromator entrance slit. Moreover, the quantum efficiency of the CCD camera (Figure 3.4) is different from the photomultiplier quantum efficiency (about 14% at 633 nm, [1]).
The x ray irradiation setup is described in section 4.1.

5.2 Results and discussion

In the following, L_i and q_{out} are normalized to the integrated signal of the PTW ionization chamber, Ic, used for normalization of the x ray tube delivered dose. L_i is calculated over pictures with bk1 background subtracted (section 4.4)[kk].
In this chapter, the ratio L_i / q_{out}, named *number of photons per secondary electron*, is also studied as a function of GEM holes, Ar/CF_4 ratio and gas mixture purity. When considering L_i / q_{out}, the light production process is separated from the electric signal formation (because it is normalized to the number of electrons produced during the gas avalanches). Therefore, this ratio is useful for understanding the light production process and so the basic detector operation. As suggested by Fraga [4], the number of photons per secondary electron is a good parameter to compare relative scintillation

[kk] For x ray experiments, the magnitude of bk3 was found to be similar to bk1 magnitude.

yields if the geometry of the setup is kept unchanged. However, for studying the dosimetric properties of the detector L_i / q_{out} can not be used, because, for example, it is constant as a function of delivered dose.

5.2.1 Scintillating GEM detector gain

In Figure 5.1, a gain curve for small hole GEMs is shown. The gain ($G = q_{out}/q_p$) was measured for Ar + 8 % CF_4 as a function of (ΔV_{GEM1} + ΔV_{GEM2}) with $\Delta V_{GEM1} = \Delta V_{GEM2}$, while keeping the x ray tube settings fixed. The detector works with a typical gain value of $2 \cdot 10^4$ under x ray irradiation. If it is operated in a particle beam (chapters 6, 7, 8) the gain drops because the maximum stable operating voltage value is reduced. This GEM detector behaviour is known, e.g. see [2,3].

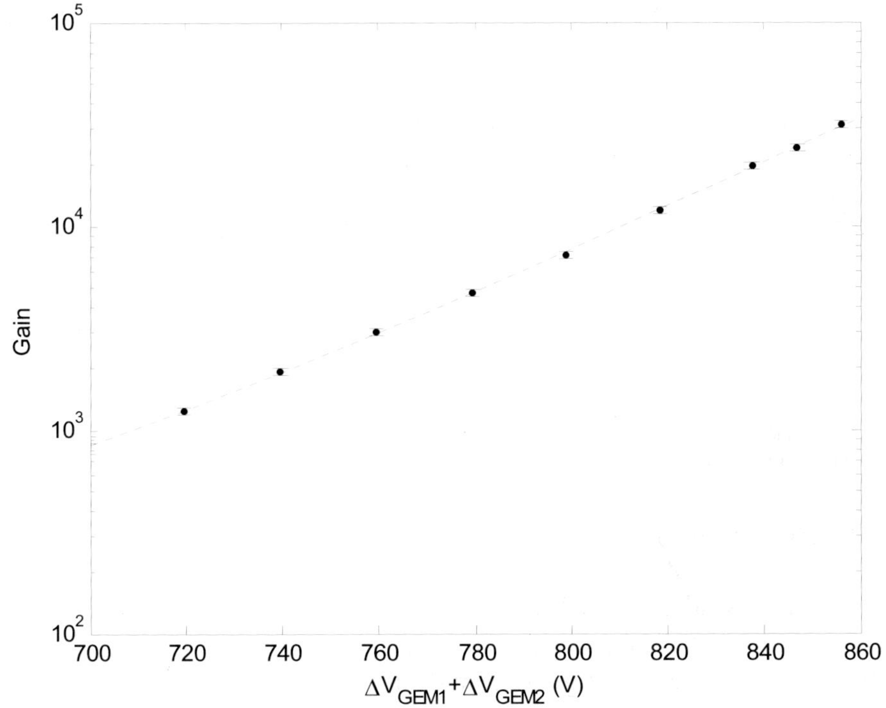

Figure 5.1 Gain curve for small hole GEMs measured as a function of $\Delta V_{GEM1} + \Delta V_{GEM2}$, with $\Delta V_{GEM1} = \Delta V_{GEM2}$.

In Table 5.1, scintillating GEM detector gain values corresponding to the used voltages in respectively x ray, proton, alpha and carbon ion beam are listed.

Beam / GEMs	Ar/CF_4	$\Delta V_{GEM1} + \Delta V_{GEM2}$ (V)	Gain
X ray small holes	92/08	425+425 = 850	24260 ± 485
X ray big holes	92/08	425+425 = 850	10444 ± 209 [ll]
Proton small holes	92/08	365+365 = 730	1556 ± 31
Alpha big holes	96/04	350+300 = 650	445 ± 9 [mm]
^{12}C small holes	92/08	350+340 = 690	749 ± 15

Table 5.1 Scintillating GEM detector gain values calculated using data of Figure 5.1 for the maximum GEM voltages used under respectively x ray, proton, alpha particle and carbon ion irradiation.

5.2.2 Relation between L_i and q_{out}

In the upper part of Figure 5.2, L_i values are shown as a function of q_{out}. q_{out} was varied by changing ΔV_{GEM1} and ΔV_{GEM2}, while keeping the x ray tube settings fixed. The data are for small hole GEMs and Ar + 8 % CF_4. L_i is linearly related to q_{out}. As explained in Chapter 3, the observed scintillating GEM detector light is produced during the gas multiplication processes in the GEM holes, to which q_{out} is directly related. In the lower part of Figure 5.2, the relative residuals[nn] of the least square linear fit to the data points are shown; Table 5.2 presents the coefficients of the fit, m and z.

[ll] The gain value of GEMs having big holes was measured in a different detector chamber then the one used for measurements with small hole GEMs. The big hole GEMs gain is of the same order of magnitude of the one reported in [4].
[mm] The gain value is for small hole GEMs, even if the experiments were performed with big hole GEMs.

[nn] Relative residuals are defined here as $(\text{relative residual})_i = \left(\frac{y_i - (m \cdot x_i + z)}{y_i} \right)$ where y_i is the experimental data measured at corresponding x_i, while m and z are the coefficients of the least square linear fit, and i=1…n, n is the data points number.

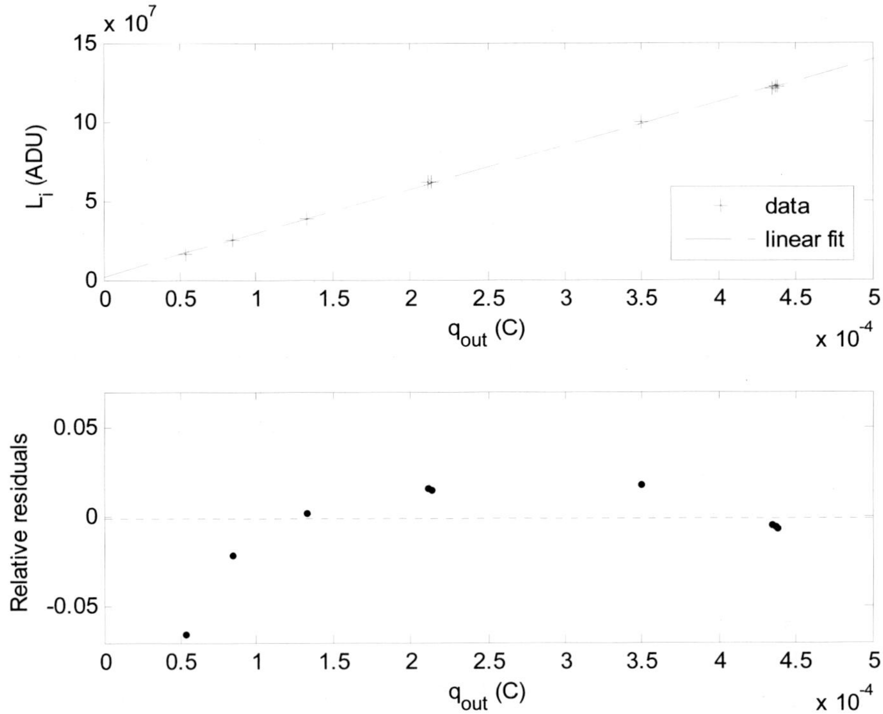

Figure 5.2 Upper graph, L_i as a function of q_{out}. Lower graph, residuals of the interpolating line, calculated with a least square linear fit.

$L_i = m\cdot q_{out}+z$	
$m \pm \sigma_m$	$(2.79 \pm 0.02)\cdot 10^{11}$ (ADU/C)
$z \pm \sigma_z$	$(1.02 \pm 0.30)\cdot 10^{6}$ (ADU)

Table 5.2 Coefficients of the least square linear fit of data shown in Figure 5.2

5.2.3 "Start up effect"

In Figure 5.3, q_{out} (upper graph) and L_i (lower graph) are shown as a function of measurement time for small hole GEMs. The data, measured in Ar + 8 % CF_4, were taken in succession for fixed x ray tube settings and fixed voltages across the GEMs.
Both detector outputs decrease till an equilibrium value is reached and their decrement is about the same, ~ 10%. The process is named here "start up effect". Gain variations with time are a known phenomenon in GEM based detectors, as mentioned in section 3.3. The time interval needed to reach a stable output, in this particular case about 15 minutes, depends on the dose

delivered to the detector sensitive volume per beam shot, on the beam intensity, on the type of GEM hole and probably on the detector internal electric field configuration.

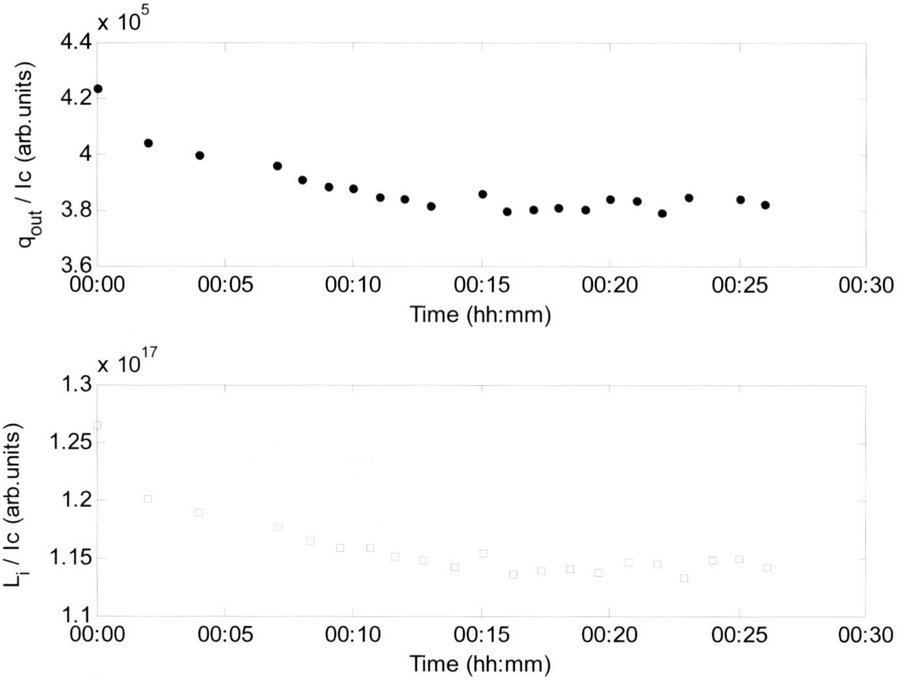

Figure 5.3 q_{out} (upper graph) and L_i (lower graph) as a function of measurement time. The data are for small hole GEMs.

The "start up effect" influences the reproducibility of the scintillating GEM detector output and therefore, its performance as a dosimeter. A solution to this problem could be to quantify how much deposited energy is needed to reach q_{out} and L_i equilibrium values and irradiate the detector, just before its usage, according to the required deposited energy. However, this operation should be repeated every time the irradiation conditions and/or the detector configuration are changed.

In Figure 5.4, L_i / q_{out} for the data represented in Figure 5.3 is shown as a function of time. On the contrary of the q_{out} and L_i decrease, the ratio varies less than 1% in the first 15 minutes.

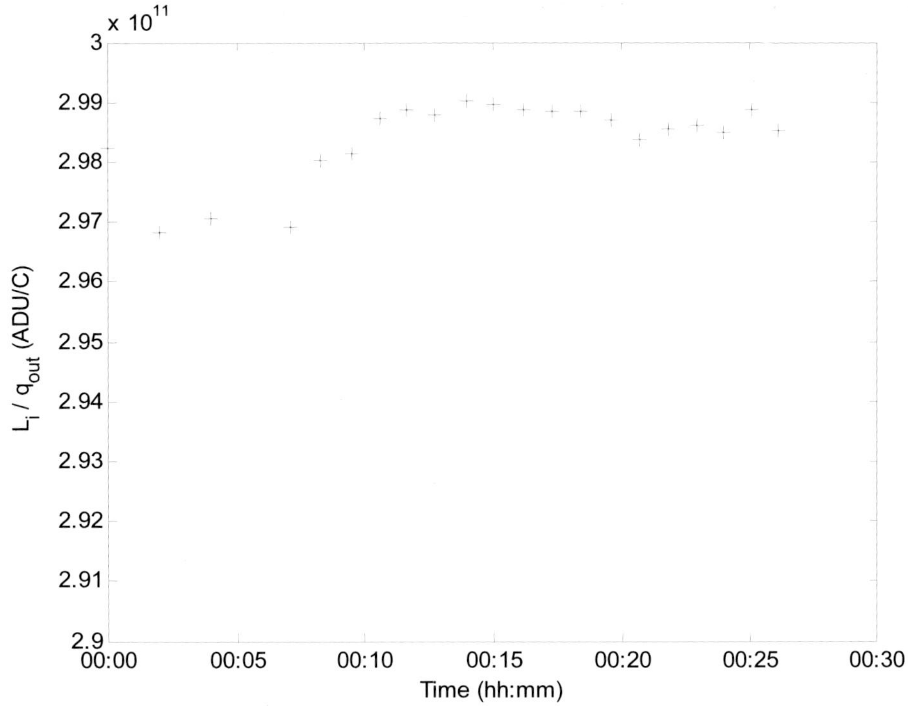

Figure 5.4 Ratio of L_i and q_{out} of the data represented in Figure 5.3 as a function of time.

5.2.4 Pressure dependence correction

The influence of pressure variations on the scintillating GEM detector outputs is not negligible [5,6] and has to be considered, especially when comparisons are made among q_{out} or L_i values measured at different times[oo]. In order to have a "calibration" curve by means of which the detector outputs can be compensated for pressure variations, we recorded q_{out} and L_i at different pressure values. The pressure was varied in the range 990-1028 mbar by closing/opening in steps a needle valve, and it was monitored by means of the pressure sensor connected to the detector chamber (section 3.1). During this experiment, the gas mixture was set to Ar + 8 % CF_4 and all the measurement conditions but the pressure were kept fixed.

[oo] Also the temperature variations should be considered but in the data comparisons done within this work, the temperature was almost constant among the different measurements. Therefore, a temperature correction curve was not performed.

In Figure 5.5, q_{out} (upper graph) and L_i (lower graph) are represented as a function of the pressure for small hole GEMs, together with an exponential fit.

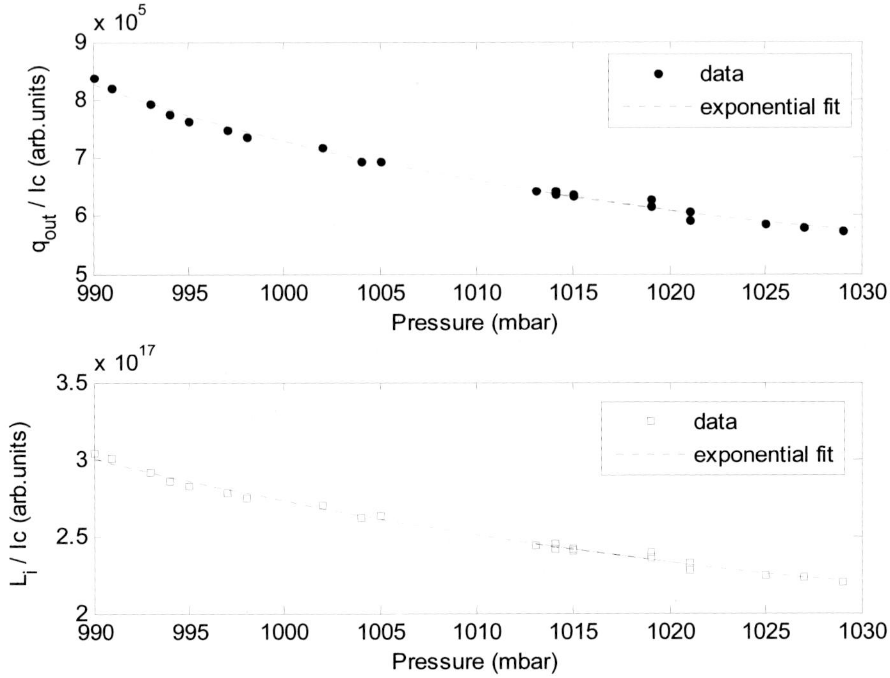

Figure 5.5 q_{out} (upper graph) and L_i (lower graph) as a function of the pressure for small hole GEMs.

The two scintillating GEM detector outputs decrease with increasing the pressure. An exponential function, such as $A + B \cdot e^{-p/C}$, was chosen for the fit because of the data trend. Table 5.3 summarizes q_{out} and L_i fit parameters[pp].

Fit parameters	A	B	C
q_{out} / Ic (arb. units)	$4.85 \cdot 10^{10}$	$1.03 \cdot 10^{25}$	29.69
L_i / Ic (arb. units)	$1.74 \cdot 10^{17}$	$1.44 \cdot 10^{28}$	38.89

Table 5.3 q_{out} and L_i exponential fit parameters. The exponential fit function is $A + B \cdot e^{-p/C}$.

[pp] The quality of the fit was verified performing a χ^2 test with 0.05 significant level of acceptance.

The fit parameters of Table 5.3 are used in the following to correct data for pressure variations. One pressure is chosen as a reference, p_{ref}, and the data that must be compared are renormalized to this p_{ref} value.
The uncertainty introduced by this pressure correction was evaluated comparing q_{out} (or L_i) values[qq] corrected for the pressure using two different series of fit parameters obtained from two calibration curves measured separately. It was found that the biggest difference between q_{out} and L_i pressure corrected values by means of the two fit parameters series was respectively of about 1.2 % and 1.3 %. These differences are considered as the uncertainties of the pressure correction procedure.
It must be stressed that the exponential fit $A + B \cdot e^{-p/C}$ is a calibration curve; it is not based on a physical model. Therefore, the fit parameters of Table 5.3 can only be used to find pressure correction factors in the studied pressure range, 990-1028 mbar. The scintillating GEM detector outputs were monitored in this pressure range because it was observed that the atmospheric pressure in the laboratory was on average changing within this range.

5.2.5 Some Ar/CF_4 emission spectrum features

5.2.5.1 Ar/CF_4 emission spectrum as a function of GEMs voltage

The small hole GEMs Ar + 8 % CF_4 emission spectrum shape is independent of the voltage supplied across the GEMs, with $\Delta V_{GEM1} = \Delta V_{GEM2}$, as can be seen in the upper graph of Figure 5.6. The spectrum area increases with the voltage, because when the latter is increased, the gain of the detector becomes higher (Figure 5.1), and so, more light is emitted. If the emission spectrum is expressed in terms of photons per secondary electrons (light signal / q_{out}), its area does not change if the voltage is increased (lower graph of Figure 5.6). So, the ratio light signal / q_{out} is GEM voltage independent. It is convenient to express the spectrum in number of photons per secondary electrons when a comparison needs to be made among spectra measured for different GEM voltages.

[qq] The q_{out} (or L_i) values considered for the pressure correction error estimation belong to an experiment that lasted two weeks, during which the pressure varied in the range 1001-1028 mbar.

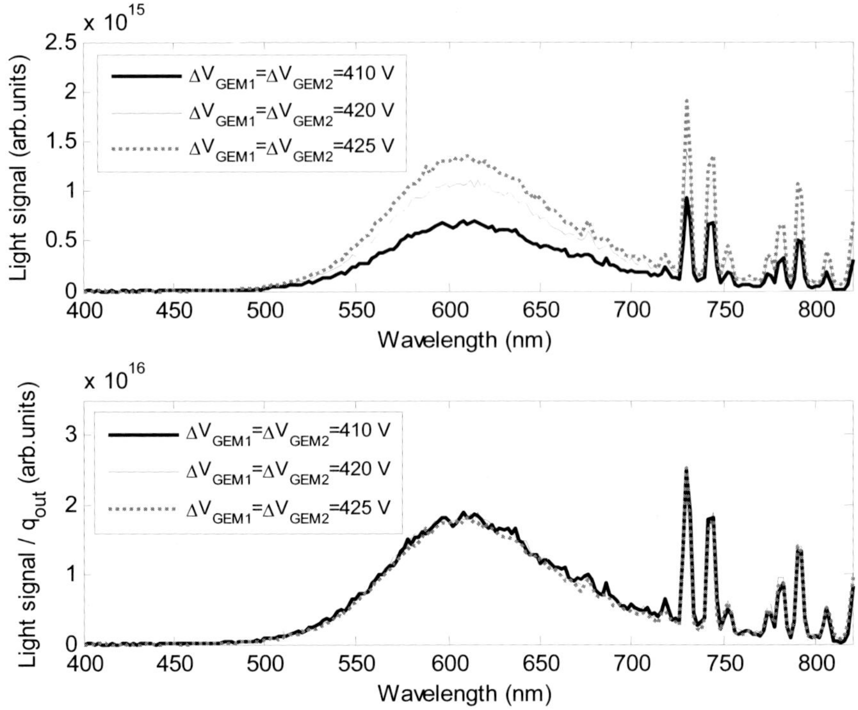

Figure 5.6 Ar + 8% CF_4 emission spectrum as a function of the voltage supplied across the GEMs, with $\Delta V_{GEM1} = \Delta V_{GEM2}$. Upper graph: on the y-axis the light signal normalized for the Ic signal is represented. Lower graph: on the y-axis the light signal divided by q_{out} is represented.

5.2.5.2 Ar/CF_4 emission spectrum as a function of x ray tube current

In order to check if the shape of emission spectrum changes with the beam rate, we collected Ar + 8 % CF_4 spectra for three different x ray tube currents, keeping the voltage difference across the small hole GEMs fixed. As shown in Figure 5.7, there are no differences among the three spectra measured with 10, 15 and 20 mA x ray tube current and constant kV [rr].

[rr] The light signal is overall higher than the spectra reported in the upper graph of Figure 5.6 because the voltage employed was higher, ~ 430 V across each GEM.

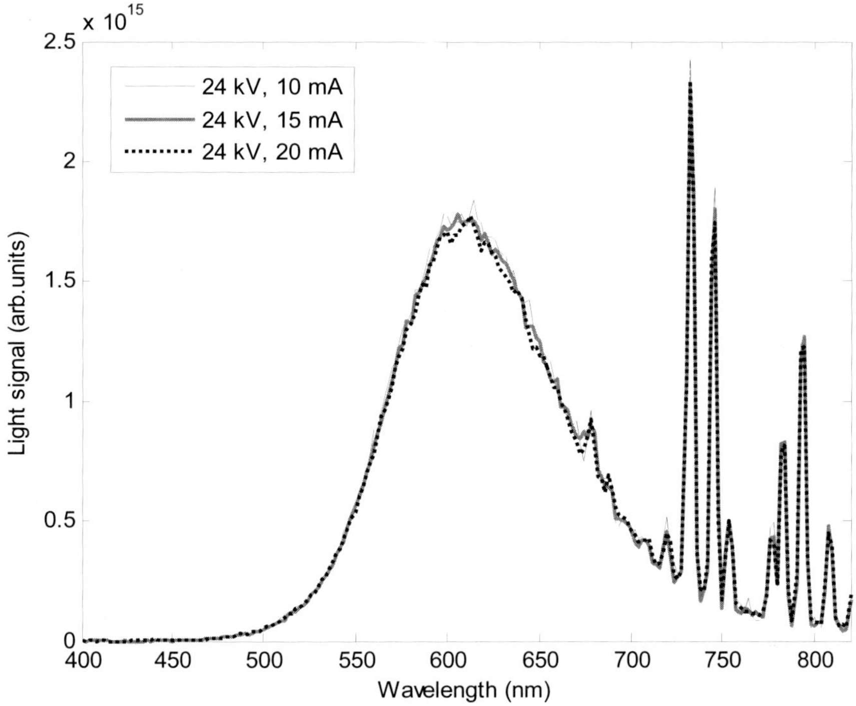

Figure 5.7 Ar + 8 % CF_4 spectra for three different x ray tube currents, 10, 15 and 20 mA. The voltage across the small hole GEMs is the same for the three spectra.

5.2.6 GEM hole shape

We investigated the response of the detector as a function of the GEM hole size and pitch. q_{out} and L_i were measured for GEMs having small and big holes (section 3.1) in Ar + 6% CF_4 as a function of (ΔV_{GEM1} + ΔV_{GEM2}) with $\Delta V_{GEM1} = \Delta V_{GEM2}$. Before performing the experiment, E_d and E_t were optimized for both kinds of GEM holes.
As can be seen in the upper graph of Figure 5.8, at larger (ΔV_{GEM1}+ΔV_{GEM2}) values q_{out} measured with GEMs having small holes is slightly higher than q_{out} measured for big holes. The difference in L_i values is bigger than the q_{out} difference and it is independent of the supplied voltages across the GEMs, as shown in the lower graph of Figure 5.8. In Table 5.4, q_{out}, L_i and their ratios for small and big holes are summarized for data at (ΔV_{GEM1}+ΔV_{GEM2}) = 739 V.

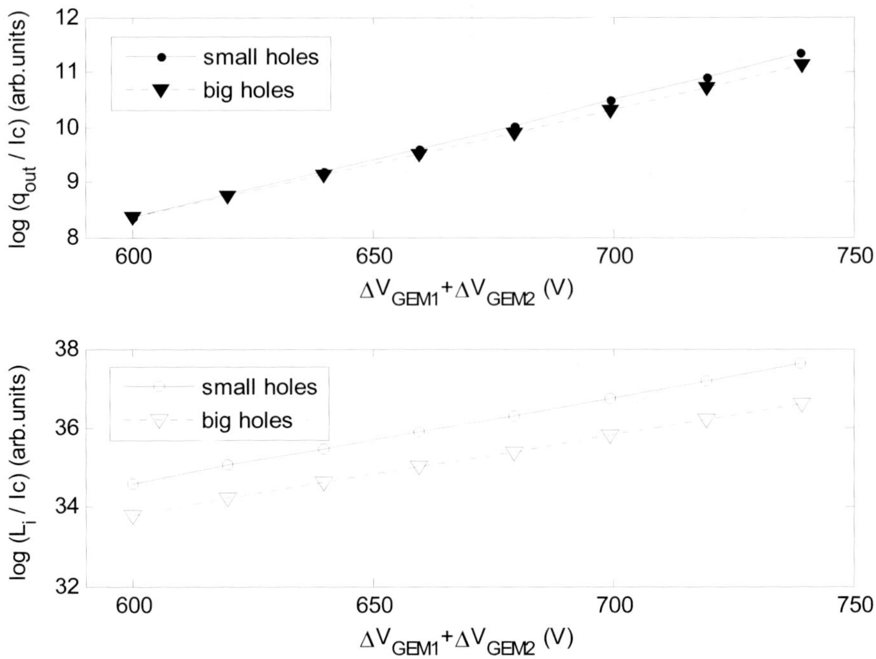

Figure 5.8 q_{out} (upper graph) and L_i (lower graph) as a function of $\Delta V_{GEM1}+\Delta V_{GEM2}$ measured for GEMs having small and big holes, for Ar + 6 % CF_4.

	q_{out} / Ic (arb.units)	L_i / Ic (arb.units)	L_i / q_{out} (ADU / C)
Small holes	$(8.62 \pm 0.17) \cdot 10^4$	$(2.19 \pm 0.05) \cdot 10^{16}$	$(2.54 \pm 0.07) \cdot 10^{11}$
Big holes	$(6.84 \pm 0.13) \cdot 10^4$	$(0.80 \pm 0.02) \cdot 10^{16}$	$(1.17 \pm 0.03) \cdot 10^{11}$
Ratio	1.26 [ss]	2.74	2.17

Table 5.4 q_{out}, L_i, and their ratio measured for GEMs having small and big holes at $(\Delta V_{GEM1}+\Delta V_{GEM2})$ = 739 V in the same detector chamber.

[ss] The q_{out} for small holes is 1.26 times the big holes q_{out}. In Table 5.1, the gain measured for small holes is 2.3 times higher than the big GEM holes. This difference is due to the fact that in the first case the same chamber was used and the two couple of GEMs were swapped in between the two measurements, while in the second case the measurements were performed using two different detector chambers. Therefore, the measurements were performed for the same level of impurities. While in the second case, the level of impurities was different for the small and big holes GEMs. According to the results presented in section 5.2.8.2 and 5.2.8.3, the different level of impurities can explain the difference between the two slightly different small holes/big holes ratios.

The higher q_{out} obtained with small hole GEMs is in agreement with Figure 13 of [6], in which it is shown that the GEM detector gain increases reducing the hole diameter. The L_i small/big holes ratio is by a factor of about 2.2 higher than the q_{out} ratio. For the L_i ratio no data were found in literature. However, in [7] a higher L_i / q_{out} ratio is also measured in He/CF_4 for small hole GEMs compared to big hole GEMs.
In Figure 5.9, the Ar + 8 % CF_4 emission spectra measured for GEMs having small and big holes are shown.

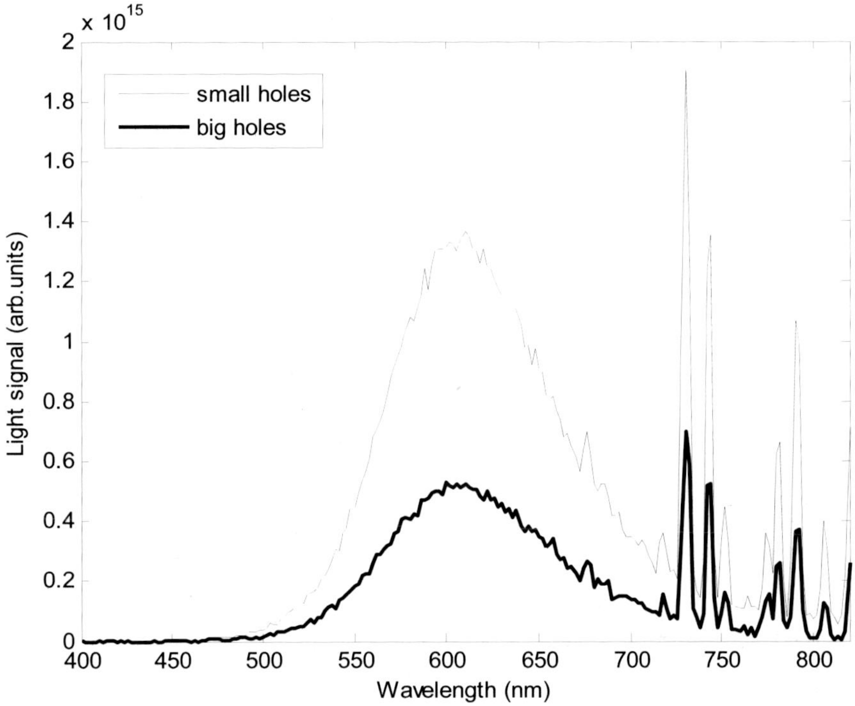

Figure 5.9 Ar + 8 % CF_4 emission spectra measured with GEMs having small and big holes. On the y-axis the light intensity normalized to the Ic signal is reported.

The two spectra present the same shape but not the same intensity. Therefore, the higher light recorded with small hole GEMs can not be attributed to a difference in emitting species.
The brighter light signal for small holes can be related to the electric field configuration along the hole axis. As represented in Figure 3.8, under the same detector gain the simulated electric field strength for small holes is higher in the hole centre compared to that in bigger holes [8]. Qualitatively, according to Figure 3.12 more intense electric field corresponds to higher

ionization probability for Ar and for CF_4. Consequently, in a stronger electric field more electrons are created and more photons are emitted.
The fact that for small holes the number of photons per secondary electron is larger than that for big holes must be related to the excitation probability. If the excitation of Ar and CF_4 is more probable than their ionization, then more photons are produced with respect to the electrons. Further studies are needed to understand better this subject.

5.2.7 Ar/CF_4 ratio

5.2.7.1 q_{out} and L_i

The gas mixtures Ar + % CF_4, with % CF_4 = 2, 4, 6, 8, 10 were investigated for small hole GEMs. For each gas mixture, q_{out} and L_i were recorded as a function of the sum of GEM voltages. The latter were increased till the maximum operating GEM voltage[tt] was reached.
In the upper graph of Figure 5.10, the q_{out} amplification curves for the gas mixtures mentioned above are represented. The highest maximum operable GEM voltage is reached in Ar + 8/10 % CF_4. However, the highest q_{out} value is measured for Ar + 8 % CF_4. These curves are in agreement with, for example Figure 1 of [9].
In the lower graph of Figure 5.10, L_i amplification curves are shown. The highest L_i value is measured for Ar + 8% CF_4, at the highest maximum operable GEM voltage. For a fixed (ΔV_{GEM1} +ΔV_{GEM2}), for example 819 V, L_i decreases when the CF_4 concentration is increased from 4 % to 10 %. The same can be concluded for q_{out}.
Figure 5.11 represents the ratio L_i / q_{out} as a function of (ΔV_{GEM1} +ΔV_{GEM2}) for the different gas mixtures investigated. For a fixed Ar + % CF_4 mixture, the number of photons per secondary electron is, as already seen in Figure 5.6, in first approximation independent of the GEM voltage and it increases when the CF_4 concentration increments from 2 to 8 %. For the latter concentration, L_i / q_{out} has a maximum. Also in [10], it is shown that the number of photons normalized to q_{out} is independent of the detector gas gain. Concluding, Ar + 8% CF_4 allows reaching the highest stable GEM operable voltage, which corresponds to the highest q_{out}, the brightest L_i and the highest L_i / q_{out}.

[tt] The maximum operable voltage is the GEM voltage above which the GEM undergoes steady discharge mode or the power supply “trips” because of an over current.

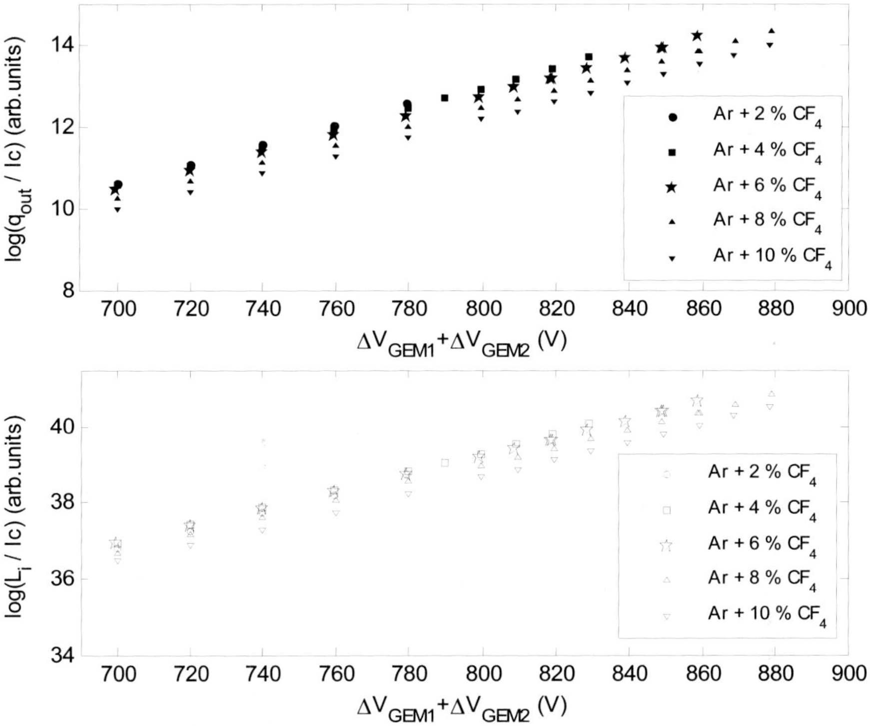

Figure 5.10 q_{out} (upper graph) and L_i (lower graph) as a function of the sum of the voltages applied to GEM_1 and GEM_2 (amplification curve), with $\Delta V_{GEM1} = \Delta V_{GEM2}$ for several gas mixtures. For Ar + % CF_4, with % CF_4 = 2, 4, 6, 8, 10. The data are for small hole GEMs.

5.2.7.2 Emission spectra

In Figure 5.12, an Ar + 4% CF_4 emission spectrum is compared to an Ar + 10 % CF_4 and Ar + 20 % CF_4 spectra. The spectra were collected for small hole GEMs. It is clearly visible that the intensity of the Ar atomic lines decreases when the percentage of CF_4 is increased from 4 to 20 % (the y-axis limits are the same for the three different graphs). On the other hand, the height of the CF_4 visible broad band is almost the same for 4%, 10 % and 20 % CF_4. According to [11], the fact that the height of the CF_4 visible broad band is almost constant for increasing CF_4 concentrations, suggests that the direct excitation of CF_4 molecules by electron impact is not the only channel leading to the CF_3* emission (section 3.6.1).

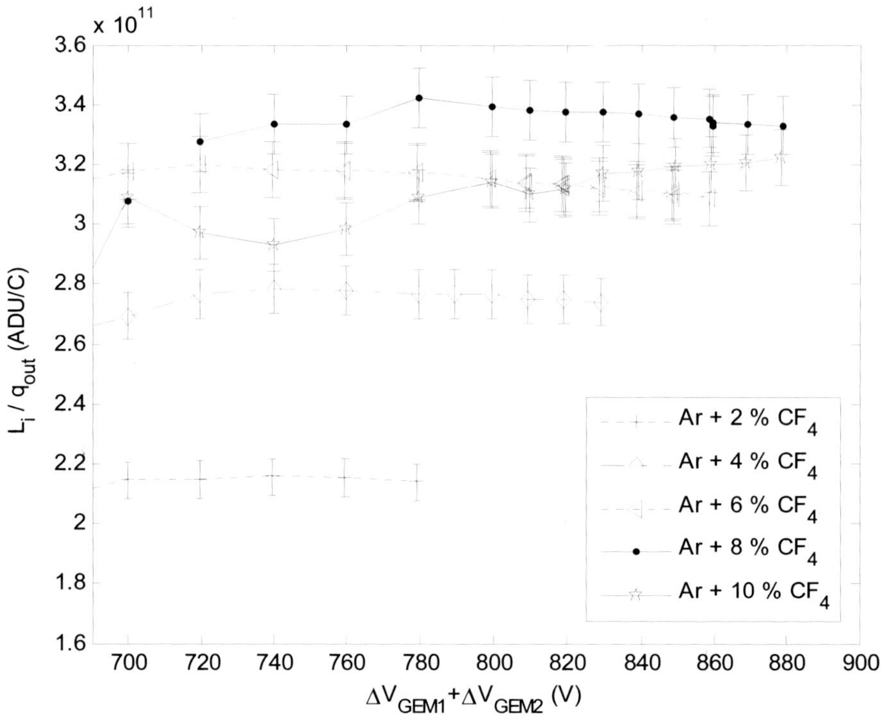

Figure 5.11 L_i / q_{out} as a function of the GEMs voltage, measured for Ar +%CF_4, with % CF_4 = 10, 8, 6, 4, 2.

An energy transfer mechanism between the excited states of argon and the dissociative electronic excited states of CF_4 may be present. Apparently, this mechanism is becoming more efficient than the Ar emission when the CF_4 quantity is increased.

Unfortunately, a quantitative[uu] comparison between the three spectra of Figure 5.12 and the integrated L_i / q_{out} values of Figure 5.11 is not possible because the latter were not recorded for Ar + 20 % CF_4. Besides, (1) they were measured with a different impurity level inside the detector chamber than the emission spectrum measurements (in between the two experiments some leaks in the detector chamber were discovered and fixed); (2) as mentioned in section 5.1, L_i the light collection angle of the lens system is

[uu] A qualitative comparison could in principle be possible. However, the intensities of the argon lines and of the CF_3^* continuum are not constant. Therefore, a deconvolution of the spectrum with the response function of the monochromator should be performed prior to the integration because continua and atomic lines are affected by the width of the setup sensitivity differently.

different from the one of monochoromator entrance slit and (3) the CCD camera quantum efficiency is different than that of the photomultiplier tube.

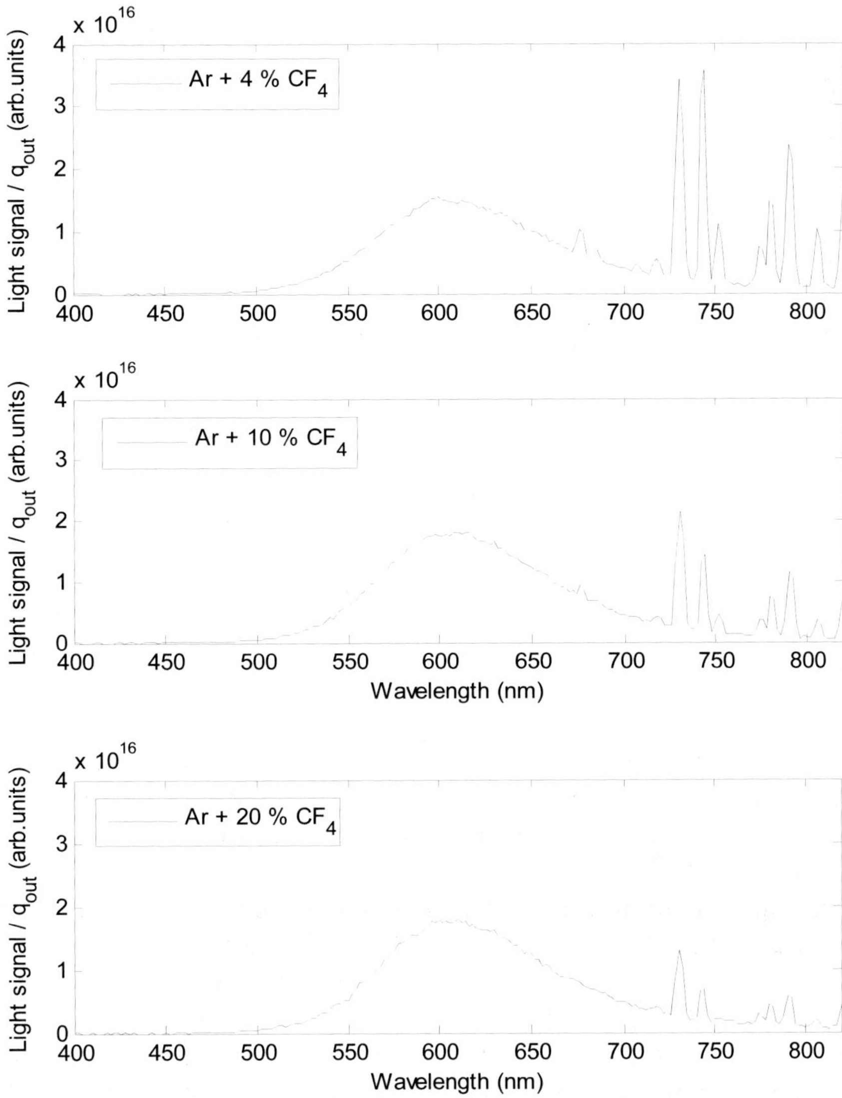

Figure 5.12 Emission spectra, of Ar + 4 %CF_4; Ar + 10 % CF_4 and Ar + 20 %CF_4. On y-axis, the ratio L_i / q_{out} is reported.

The differences between the upper and lower graph of Figure 5.12 are similar to the differences visible in the two spectra of Figure 2 of [11], measured respectively for Ar + 5% CF_4 and Ar + 67 % CF_4. In Figure 5.12, the quenching of the Ar lines when the CF_4 concentration is increased from

4 to 20% is higher than the quenching of the same lines reported in Figure 2 of [11], where the CF_4 concentration increments from 5 to 67%. Most likely this effect is due to presence of some air inside the scintillating GEM detector chamber, that acts as a quencher.

5.2.8 Gas mixture purity

5.2.8.1 Varying the initial purity of the gas mixture

The dependence of the scintillating GEM detector outputs on the initial gas mixture purity was investigated. The purity of the mixture was varied changing the Ar purity, while keeping fixed the CF_4 gas purity. q_{out} and L_i were measured in Ar + 8 % CF_4 for three gas mixture purities: 99.990 %, 99.996 %, 99.999 %. Small hole GEMs were used. In Table 5.5, an overview of the results is presented for $(\Delta V_{GEM1} + \Delta V_{GEM2}) = 850$ V.

Gas mixture purity	q_{out} / Ic (arb.units) [vv]	L_i / Ic (arb.units) [vv]	L_i / q_{out} (ADU / C)
99.990 %	$(1.78 \pm 0.03) \cdot 10^{-5}$	$(5.92 \pm 0.11) \cdot 10^{6}$	$(3.33 \pm 0.08) \cdot 10^{11}$
99.996 %	$(1.94 \pm 0.04) \cdot 10^{-5}$	$(6.55 \pm 0.15) \cdot 10^{6}$	$(3.36 \pm 0.10) \cdot 10^{11}$
99.999 %	$(1.87 \pm 0.04) \cdot 10^{-5}$	$(6.16 \pm 0.14) \cdot 10^{6}$	$(3.30 \pm 0.10) \cdot 10^{11}$

Table 5.5 q_{out} and L_i values, and their ratio as a function of the Ar + 8 % CF_4 purity. The latter was set respectively to 99.990 %, 99.996 %, and 99.999 %. The data are for small hole GEMs.

Increasing the gas mixture purity of 0.06 ‰ causes a q_{out} and L_i increment by a factor of about 1.09. When an even purer gas mixture is used (99.999 %), q_{out} is within the uncertainty compatible with the value obtained for 99.996 % purity. On the other hand, L_i slightly decreases when the purity is increased from 99.996 % to 99.999 %. Probably, this gas mixture purity is higher than the purity level inside the detector chamber. The impurities present in the chamber and gas handling system "contaminate" the initial gas mixture purity and influence the light production process, quenching it.

[vv] The signal of the Ic was in this particular experiment readout on the NI DAQ board in volts and then offline integrated. While in the other experiments reported in this chapter, the integrated signal of Ic is readout on the Keithley charge monitor in Coulomb. Therefore, the q_{out}/Ic and L_i/Ic ratios are in this case different.

The light signal is sensitive to the presence of impurities, as mentioned in section 3.4. The ratio L_i / q_{out} is independent of the initial gas mixture purity within the experimental errors in the studied range.

5.2.8.2 Closing the gas flow

The effect of closing the gas flow on q_{out} and L_i can be seen respectively in the upper graph and in the lower graph of Figure 5.13.

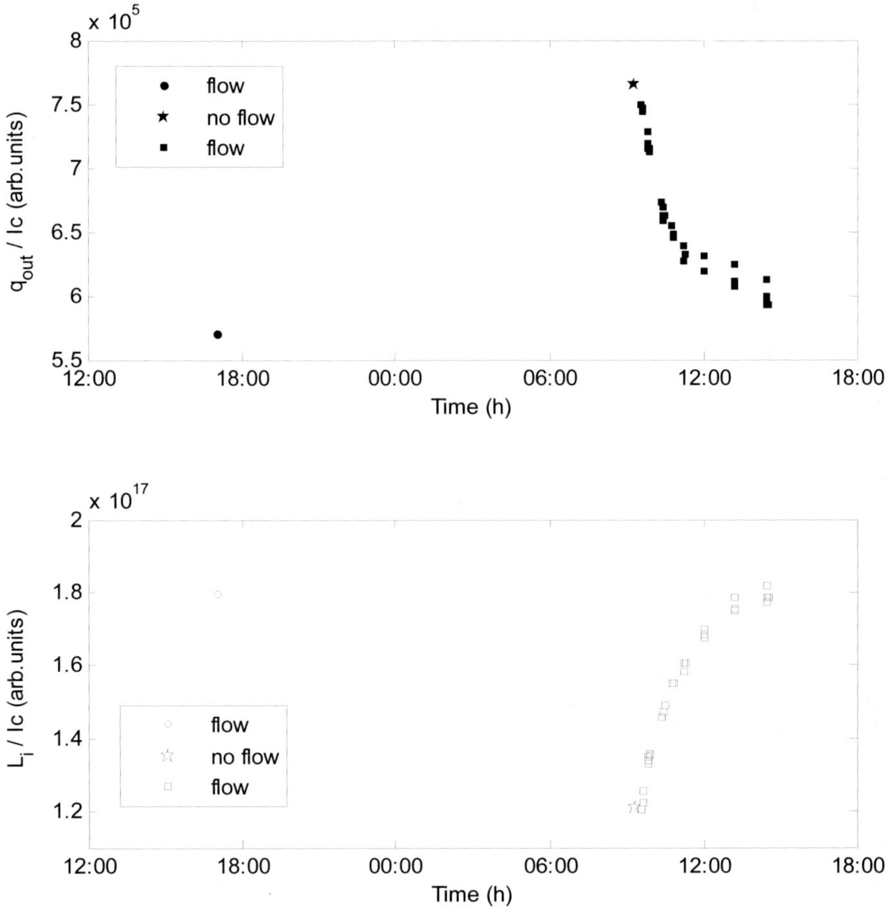

Figure 5.13 q_{out} (upper graph) and L_i (lower graph) dependence on the gas flow as a function of time. Measurement conditions: Ar + 8 % CF_4, small hole GEMs, $(\Delta V_{GEM1} + \Delta V_{GEM2}) = 850$ V.

Under the same measurement conditions (Ar + 8 % CF_4, small hole GEMs, $(\Delta V_{GEM1} + \Delta V_{GEM2}) = 850$ V), if the flow is stopped for about 16 hours q_{out} increases by a factor of about 1.3, while L_i decreases by a factor of about 1.5.

When the flow is started again, it takes roughly 5 hours for the detector outputs to reach their initial values, measured before closing the gas flow. Closing the flow implies modifying the level of impurities inside the detector chamber. The mixture inside the detector chamber ages due to, for example, the presence of out gassing materials. Therefore, as already said since the light production process is depending on the pureness of the gas mixture [12], a decrease in L_i is expected when there is no gas flow in the detector chamber. On the other hand, the increase of q_{out} when the gas flow is closed was not expected. The increase of the detector gain in presence of no gas flow could be explained by the reduction of the CF_4, quencher gas. According to Figure 5.10, if the CF_4 concentration decreases, q_{out} increases for a fixed ($\Delta V_{GEM1} + \Delta V_{GEM2}$). The CF_4 decrement could be caused by chemical reactions of the former with some impurities present in the detector chamber. This hypothesis should be verified by a gas chromatography analysis of the Ar/CF_4 mixture; in particular, it should be understood which are the chemical reactions that reduce the CF_4 concentration. As can be seen in Figure 5.14, the number of photons per secondary electron decreases by a factor of two when the flow is stopped

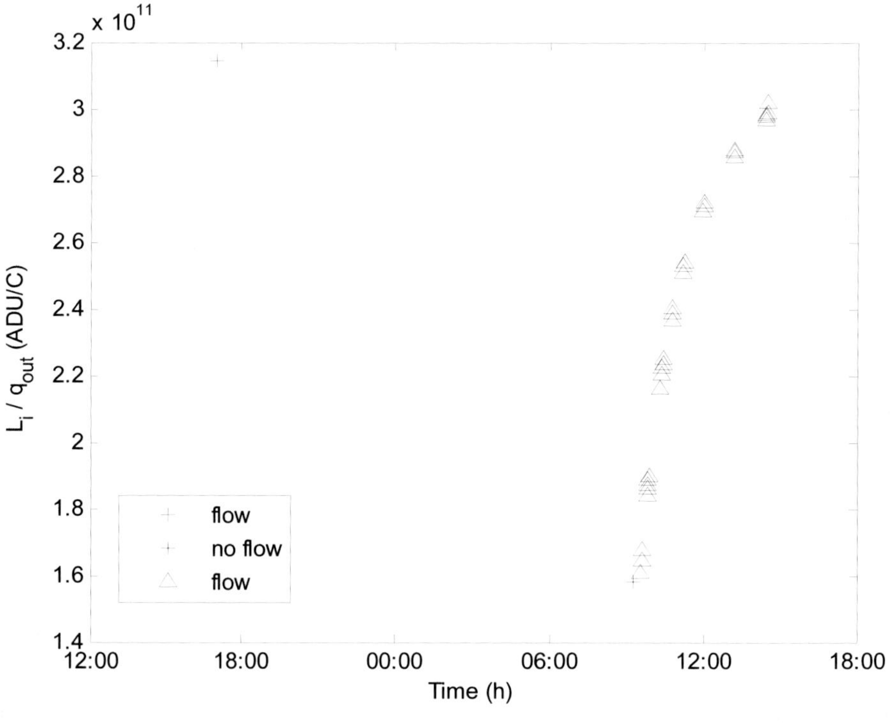

Figure 5.14 L_i / q_{out} dependence on the gas flow as a function of time. Measurement conditions: Ar + 8 % CF_4, small hole GEMs, ($\Delta V_{GEM1} + \Delta V_{GEM2}$) = 850 V.

5.2.8.3 Varying the gas flow rate

In Figure 5.15, the dependence of q_{out} (upper graph) and L_i (lower graph) on the gas flow rate is depicted as a function of the measurement sample number. The data are for Ar + 8 % CF_4, small hole GEMs, and $(\Delta V_{GEM1} + \Delta V_{GEM2})$ = 780 V. q_{out} increases by a factor of about 1.08 when the flow rate is reduced. q_{out} values measured for 4 times smaller flow rate than usual are compatible within the uncertainties with the values recorded for a flow rate of 4.5 l/h. When the flow rate is halved, L_i decreases. Overall, the reduction of the flow rate has a bigger effect on L_i compared to the electric output. To a lesser extent, reducing the flow rate affects the GEM detector outputs in the same way as closing the flow.

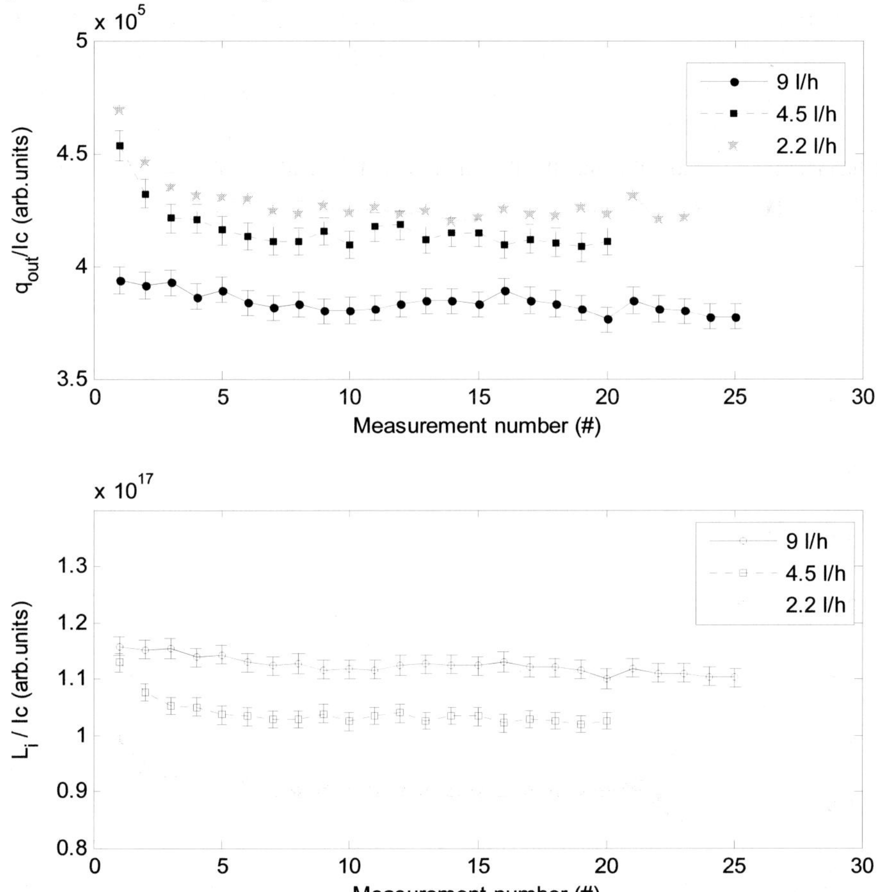

Figure 5.15 Dependence of q_{out} (upper graph) and L_i (lower graph) on the gas flow rate is depicted as a function of the measurement number. The data are relative to Ar + 8 % CF_4, small hole GEMs, $(\Delta V_{GEM1} + \Delta V_{GEM2})$ = 780 V.

In Figure 5.16, the ratio L_i / q_{out} is represented for the three different investigated flow rate values. It reduces by a factor of about 1.2 when the flow is halved and by a factor of about 1.4 when the flow rate becomes 4 times smaller.

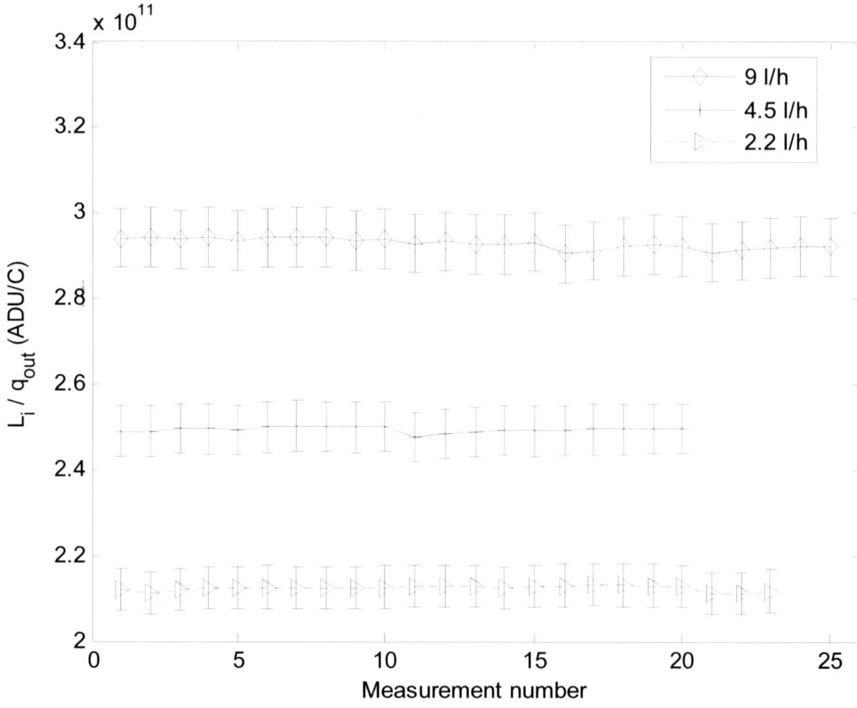

Figure 5.16 Dependence of L_i / q_{out} on the gas flow rate is depicted as a function of the measurement sample number. The data are for Ar + 8 % CF_4, small hole GEMs, (ΔV_{GEM1} +ΔV_{GEM2}) = 780 V.

5.3 Conclusions

We have found that the integrated light yield is linearly dependent on the total charge extracted from GEM_2 holes. This was expected since the photons are produced during the gas multiplication process.

q_{out} and L_i reach an equilibrium value only a certain time after the irradiation has started. The time interval is about the same for both detector outputs. This so called "start up effect" affects the detector output reproducibility. A solution to this problem could be, for a certain detector configuration and certain irradiation conditions, to quantify how much deposited energy is needed to reach q_{out} and L_i equilibrium values and irradiate the detector just before its usage according to the required deposited energy. Anyway, in anticipation of a clinical use of the scintillating GEM detector the "start up effect" should be studied in more detail.

The detector outputs must be compensated for pressure variations. In this chapter, an example of q_{out} and L_i pressure correction was presented. For a more user friendly operation of the GEM dosimeter, this kind of compensation should be implemented in the data analysis routine before displaying the detector outputs or stabilizing the detector pressure by means of a feedback system.

It was verified that the shape of the Ar/CF_4 emission spectrum is independent of (1) the voltages applied across the GEMs, (2) the x ray beam current, and (3) the GEM hole diameter. The ratio between CF_4 and Ar spectrum area changes when their concentrations are varied.

With smaller diameter GEM holes, a brighter light signal and a higher electric output are measured than with wider holes. This results can be explained by the electric field configuration along the hole axis. Under the same detector gain, the simulated electric field strength for small holes is higher in the hole centre compared to that for bigger holes. Qualitatively, more intense electric field corresponds to higher ionization probability for Ar and for CF_4. Consequently, more electrons are created and more photons are emitted. The difference in light signal found for small and big holes is bigger than the difference found in the output charge. The Ar + 8 % CF_4 emission spectra measured for GEMs having small and big holes show the same shape. Therefore, the higher light recorded with small hole GEMs can not be attributed to different emitting species. The fact that for small holes the number emitted photons per secondary electron is larger than that for big holes must be related to the excitation probability. If the excitation of Ar and CF_4 is more probable than their ionization, then more photons are produced with respect to the electrons. Further studies are needed to understand better this subject.

Varying the CF_4 concentration, it was found that Ar + 8 % CF_4 allowed reaching the highest voltage across the GEMs and the highest q_{out} and L_i value.

The last part of this chapter was dedicated to the gas mixture purity dependence of the detector outputs. Firstly, the initial purity of the gas mixture was changed by varying the Ar purity. Increasing the gas mixture purity level from 99.990 % to 99.996 % causes a q_{out} and L_i increment of about 9 %. No improvement in the signal magnitude is gained when the purity in increased by a further 0.03 ‰. The ratio L_i / q_{out} is independent on

the gas mixture purity within the experimental errors. This indicates that the detector chamber and the gas handling system purity is lower than the initial gas mixture purity values investigated in this study.
Secondly, the purity of the gas mixture was varied closing the gas flow. It has been observed that q_{out} increased by a factor of about 1.3 and L_i decreased by a factor of about 1.5 when the flow was stopped for 16 hours. Closing the gas flow causes the gas inside the chamber to age due to the presence of, for example, out gassing materials. The latter pollute the gas and consequently influence the light production process. The increase of q_{out} and so of the detector gain, in absence of flow could be explained by the reduction of the quencher CF_4 concentration due to chemical reactions of the latter with some impurities present in the detector chamber. This hypothesis should be verified by a gas chromatography analysis of the gas mixture in presence and not of flow.
Finally, the gas mixture purity was varied by reducing the gas flow rate. To a lesser extent, reducing the flow rate affects the GEM detector outputs in the same way as when the flow is closed.
All in all, varying the initial gas mixture purity by 0.06 ‰ influences the scintillating GEM detector outputs in the same way. On the other hand, when the gas mixture purity is changed by closing the flow or reducing the flow rate, q_{out} and L_i behave in a different way and they do not follow each other.
According to this purity study, we can conclude that a much “cleaner” chamber is needed in anticipation of a clinical use of the detector to guarantee a more reproducible detector response.

5.4 References

[1] http://sales.hamamatsu.com/assets/pdf/parts_R/R943-02.pdf
[2] S.Bachmann et al., Discharge studies and prevention in the gas electron multiplier (GEM), *Nucl. Instrum. and Methods*, A 479, 2002, 294.
[3] A.Bressan et al., Development of high gain GEM detectors, *IEEE Trans. on Nucl.Science,* 47, 2000, 2070.
[4] F.Fraga, Performance of a tracking device based on the GEM scintillation, *IEEE Transactions on nuclear science*, 49, 2002, 281.
[5] M. C. Altunbas et al., Aging Measurements with the Gas Electron Multiplier, N*ucl. Instr. and Methods*, A 515, 2003, 249.
[6] A. Kozlov et al., Development of a triple GEM UV-phoaton detector operated in pure CF_4 for the PHENIX experiment, *Nucl. Instr. and Methods*, A 523, 2004, 345.

[7] F.Fraga et al., CCD readout of GEM-based neutron detectors, *Nucl. Instrum. Methods*, A 478, 2002, 357.
[8] Bachmann et al., Charge amplification and transfer processes in the gas electron multiplier, *Nucl. Instrum. and Methods*, A 438, 1999, 376.
[9] M.Deptuch, T.Z. Kowalski, Gas multiplication process in mixtures based on Ar, CO_2, CF_4 ,*Nucl. Instrum. and Methods*, A 572, 2007, 184.
[10] M.Fraga et al., Pressure dependence of secondary NIR scintillation in Ar and Ar/CF_4, *IEEE Transactions on Nuclear Science*, 48, 2001,330.
[11] M. Fraga et al., The GEM scintillation in He–CF_4, Ar–CF_4, Ar–TEA and Xe–TEA mixtures, *Nucl. Instrum. and Methods*, A 504, 2003, 88.
[12] F. Fraga et al., Quality control of GEM detectors using scintillation techniques, *Nucl. Instrum. and Methods*, A 442, 2000, 417.

Chapter 6

Scintillating GEM detector dosimetric properties in a proton beam

In this chapter, the experiments performed in a proton beam are discussed. The aim of these experiments was to verify if the scintillating GEM detector can be used as a dosimeter. Therefore, it was verified if its response is reproducible, uniform, and linear with the dose and the field size, independent on the dose rate. The energy dependence of the detector output was also studied. The spatial and time responses have been measured.
The results discussed belong to two different measurement periods: during one of them, big hole GEMs were mounted in the detector, flushed with Ar + 6 % CF_4; during the other, small hole GEMs and Ar + 8 % CF_4 were employed. In each section, the type of GEMs and the gas mixture used are specified.
As already mentioned, for the scintillating GEM detector the intensity distribution of the measured light spot is proportional to the 2D hadron dose distribution. By placing a water bellows phantom in front of the detector, with respect to the beam direction, and varying the water thickness in steps, from zero up to beyond the hadron range, a 3D dose distribution can be reconstructed.

6.1 Experimental setup

The detector has been irradiated in a steady 150 MeV proton beam of the AGOR cyclotron at Kernfysisch Versneller Instituut, Groningen, the Netherlands [1]. A schematic of the beam line and the experimental setup is given in Figure 6.1, while in Figure 6.2 a photograph is shown.
The scatter foil consists of a 1.44 mm thick flat lead foil. For the experiments that required a homogeneous irradiation field (section 6.3.9, 6.3.10), a second inhomogeneous scatter foil made of tungsten was added 35 cm behind the first one [2,3]. An air-filled parallel plate ionization chamber is used as a beam monitor to register the beam intensity in MU/s. The MU (Monitor Unit) has been calibrated in dose to water at the

scintillating GEM detector location by means of a PTW23343 Markus chamber.

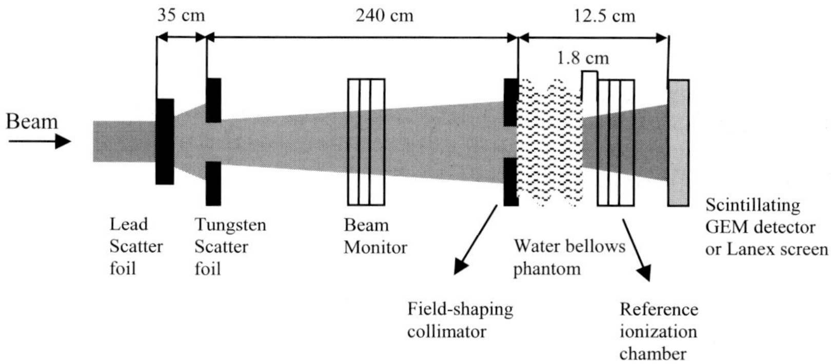

Figure 6.1 Schematic of the proton beam line and the experimental setup seen from above.

1MU is equivalent to (3.00 ± 0.09) $\cdot 10^{-6}$ Gy if both scatter foils are inserted in the beam line, while 1MU = (5.00 ± 0.15) $\cdot 10^{-6}$ Gy when only the lead scatter foil is present. The dose rate was varied by changing the proton current in the accelerator source. Experiments were performed in the dose rate range 1 – 31 Gy/min. The field–shaping brass collimator determines the beam spot size at the entrance of a water-filled bellows phantom. The field size was varied by appropriate field-shaping collimator shapes. Field sizes ranged from 1 × 10 mm^2 to 20 × 20 mm^2, and for circular collimators radii from 2.5 mm up to 40 mm were used. The beam energy was degraded by means of the water-filled bellows phantom, whose thickness can be varied in steps, of ≥ 0.05 mm, from zero up to beyond the proton range. The minimum bellows phantom water depth (0-wd) leaves 3 cm of plastic of the phantom in the beam. The charge IC-q of a second air filled parallel plate ionization chamber (IC) has been used as a reference. As can be seen in Figure 6.2, the scintillating GEM detector was placed behind the IC with respect to the beam direction.

Experiments were performed either with small hole GEMs, Ar + 8 % CF_4 and $\Delta V_{GEM1}=\Delta V_{GEM2}$=365 V, or big hole GEMs, Ar + 6 % CF_4 and $\Delta V_{GEM1}=\Delta V_{GEM2}$=370 V. For a Bragg curve measurement, a picture is taken with the water filled phantom at the minimum thickness, and at the same time I_{out} is sampled. Then, the thickness of the phantom is increased in steps up to a value larger than the proton range, and for each step a complete measurement is performed. For each water layer, L_i is obtained by integrating the picture pixels over the same region of interest.

During the measurements with the Lanex scintillating screen, the latter was fixed to a holder and the holder mounted on the plastic tube (black tube in Figure 6.2), supporting the mirror and the CCD camera, instead of the scintillating GEM detector. The holder was fixed in such a way that the Lanex screen was at the same distance from the CCD camera as GEM_2.
Due to energy losses in the beam line components, air, and the water phantom frame, the effective proton energy at the minimum water thickness was 121 MeV if both scatter foils were inserted in the beam line, and 127 MeV if only the first one was present.

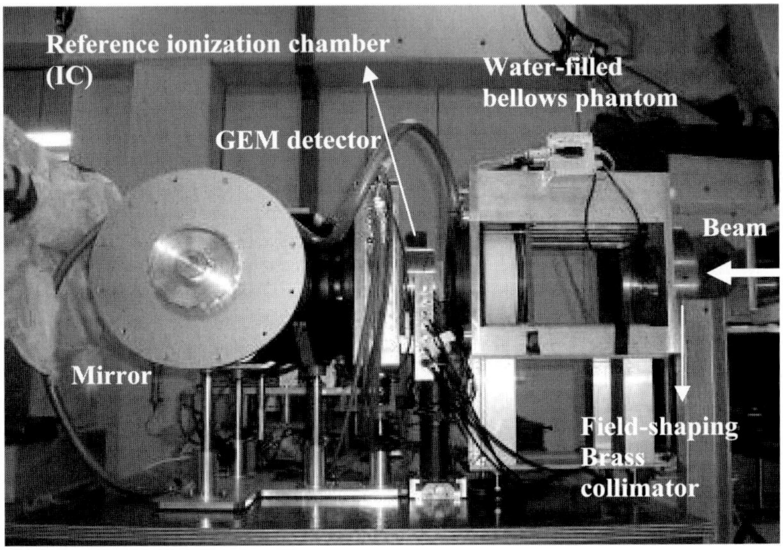

Figure 6.2 Experimental setup during proton measurements.

6.2 Nomenclature

In the following, *0-wd* indicates the minimum bellows phantom water depth; *Bp-wd* indicates the Bragg peak depth, which has about 3 times higher ionization density, or LET, than 0-wd. Some experiments were performed at both water depths to check if the ionization density increment affected the detector operation. q_{out} is defined in section 4.2; L_i is defined in section 4.3 and it is calculated over pictures with bk3 background subtracted. The q_{out} and L_i uncertainties are defined respectively in section 4.6.1 and 4.6.2. *PPR* indicates the peak to plateau ratio of a depth dose curve. The PPR is defined as the ratio of a Bp-wd measurement and a 0-wd measurement. The PPR uncertainty is calculated according to the error propagation formula.

6.3 Results and discussion

6.3.1 Light signal brightness

In Figure 6.3, the integrated GEM light signal, expressed in ADU/(pixel·Gy), as a function of the sum of the GEMs voltages is compared to the scintillating Lanex screen signal intensity. The light intensity values are for small hole GEMs and Ar + 8 % CF_4. The experiment was performed delivering 10 Gy, at 30 Gy/min, for a circular 20 mm^2 field size and at 0-wd.

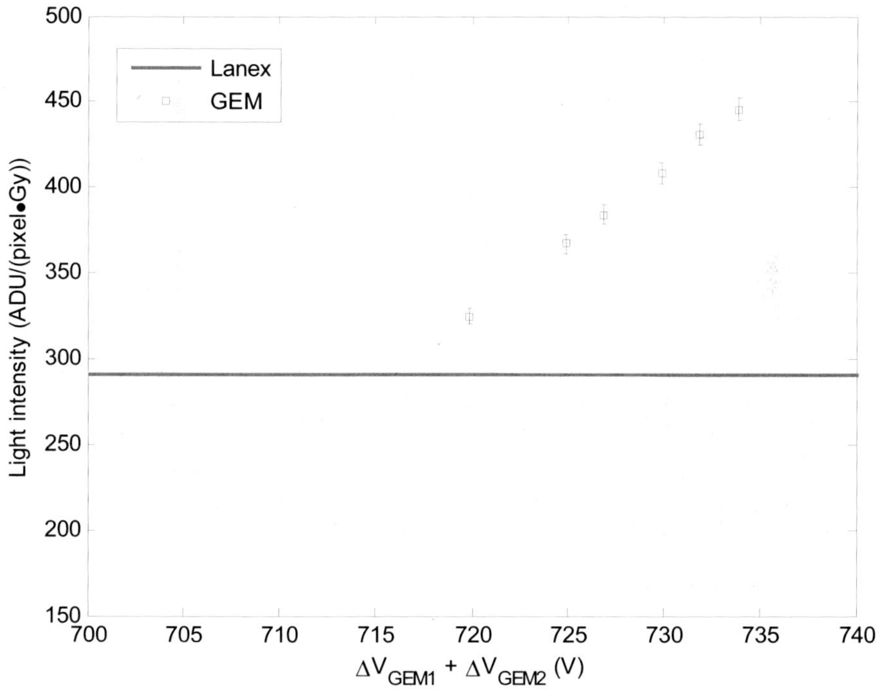

Figure 6.3 Integrated GEM light signal, in ADU/(pixel·Gy), as a function of the sum of the GEMs voltages, compared to the scintillating Lanex screen intensity.

The GEM light signal is brighter than the Lanex screen signal if $(\Delta V_{GEM1}+\Delta V_{GEM2}) > 716$ V. Most of the measurements reported in the following were performed at $(\Delta V_{GEM1}+\Delta V_{GEM2}) = 730$ V. At this tension, the GEM light signal is about 1.4 times higher than that of the Lanex screen.

6.3.2 Response reproducibility

The detector response reproducibility was evaluated using measurements performed at 0-wd under the same measurement conditions (GEMs with big holes, Ar + 6 % CF_4, 11 × 12 mm^2 field size) at different moments during the beam time period. As can be seen in Figure 6.4, the q_{out} and L_i reproducibility over two days is respectively 4 and 5%. The causes of these variations with time are not yet fully understood. Pressure and beam fluctuations are excluded because the data were compensated for them. Charge up effects could be responsible for such differences.

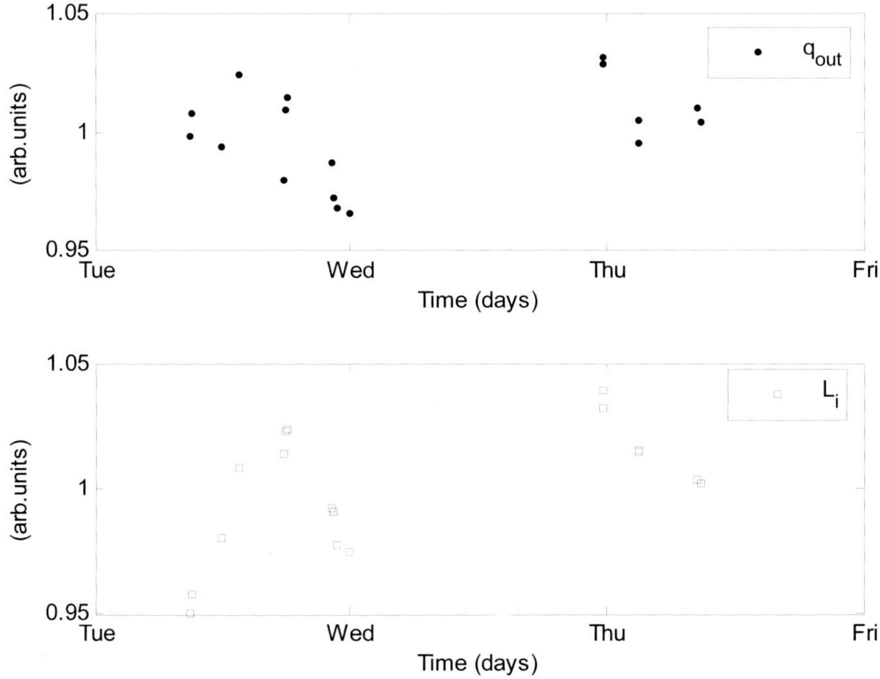

Figure 6.4 q_{out} (upper graph) and L_i (lower graph) as a function of time. The data were measured under the same measurement conditions. On the y-axis, the ratios of the single data values and the average of all the data values are reported.

6.3.3 Dose linearity

In the upper graph of Figure 6.5, L_i is shown as a function of dose. The latter was varied by changing the amount of delivered MU. The dose rate was kept fixed to 17.5 Gy/min and the field size was 11 × 12 mm^2. The measurements were performed with big hole GEMs, in Ar + 6 % CF_4. L_i is linear with the dose in the investigated range, as confirmed by the relative residuals of the least square linear fit. The same good dose linearity was found for q_{out}.

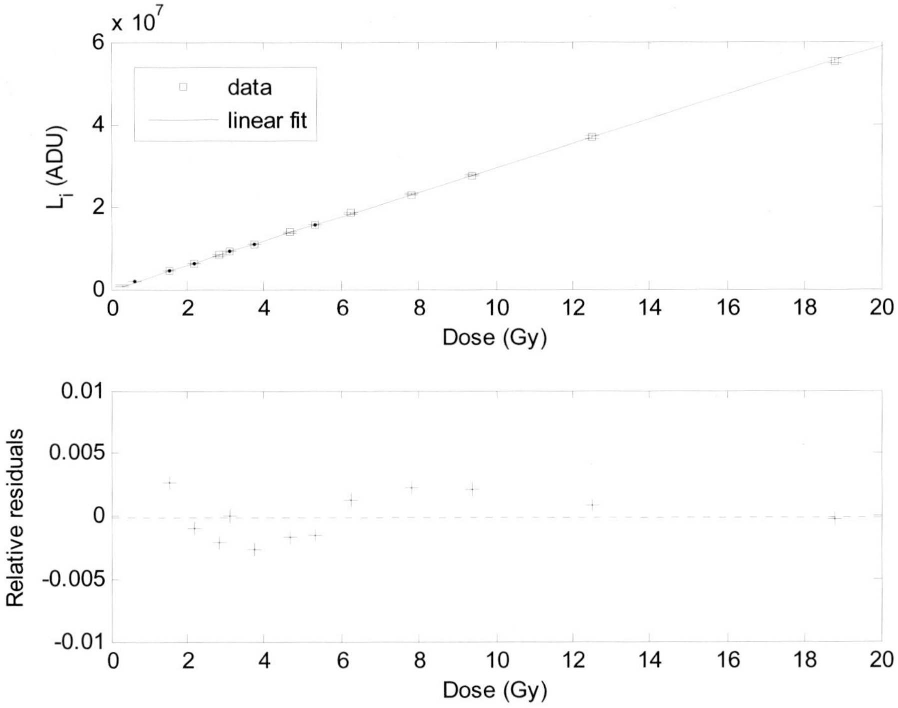

Figure 6.5 Upper graph: L_i as a function of dose in the range 1.5 – 19 Gy together with the least square linear fit. Lower graph: relative residuals of the least square linear fit.

The L_i dose linearity for doses smaller than 1.5 Gy was measured for small GEM holes and Ar + 8 % CF_4, for 20 mm^2 circular field size, and 2 Gy/min. The data are represented in Figure 6.6. According to the relative residuals, the L_i dose linearity between 0.05 and 2 Gy is within 1% except for the data at 0.5 Gy. The bigger deviation of the latter was explained by the presence of a spike in the chosen region of interest not removed by the 3-by-3 median filter, which modified the integrated light yield value. The q_{out} dose linearity was found to be within 1%.

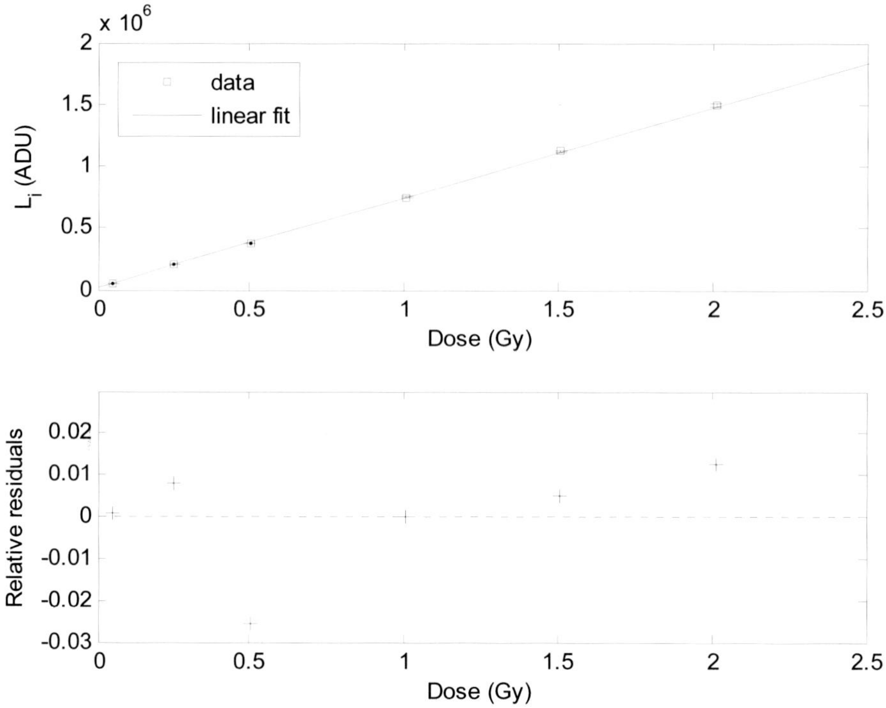

Figure 6.6 Upper graph: L_i as a function of dose in the range 0.05 – 2 Gy together with the least square linear fit. Lower graph: relative residuals of the least square linear fit.

6.3.4 Dose rate dependence

In Figure 6.7, L_i and q_{out} are shown as a function of the dose rate, together with the signal of the ionization chamber IC used as a reference. The data were measured for 11×12 mm^2 field size, 6 Gy, and big hole GEMs in Ar + 6 % CF_4. For visualization purposes, error bars in the figure are neglected. The integrated signal of IC is dose rate independent at both water depths, as expected. At 0-wd, L_i and q_{out} are constant within the ~ 2 % experimental errors in the range 1 – 16 Gy/min. The same can be concluded for the detector output measured at the Bp-wd in the range 2 – 38 Gy/min even if the random q_{out} and L_i fluctuations are bigger than those measured at 0-wd.

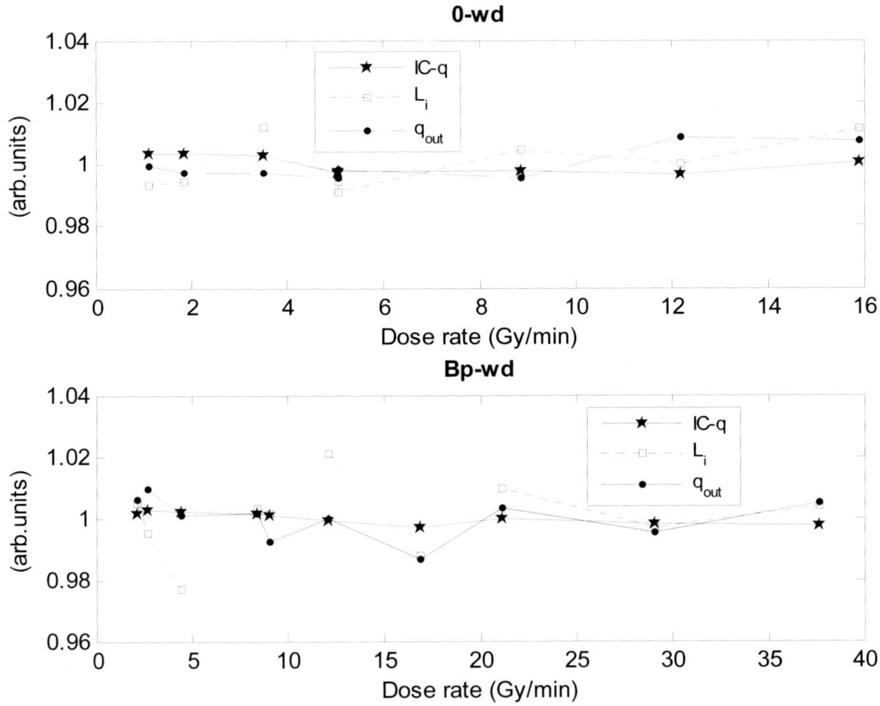

Figure 6.7 IC-q, L_i and q_{out} as a function of the dose rate for 0-wd (upper graph) and Bp-wd (lower graph). On the y-axis, the ratio of the single data values and the average of all data values are shown. Error bars, of the order of 2%, are neglected for visualization purposes.

6.3.5 Spatial response

The spatial resolution of the scintillating GEM detector was evaluated considering the light intensity profile of a picture taken with a 10 mm long and 1 mm wide field-shaping collimator. In Figure 6.9, the one-pixel wide vertical light intensity profile of the GEM detector picture taken for such a collimator (Figure 6.8) is compared to the Lanex screen intensity profile measured under the same conditions. For convenience, the latter has been adjusted (without horizontal scaling) to make the peaks of the two profiles coincide.

According to literature [4], the Lanex screen has a spatial resolution of about 0.2 mm. Therefore, the Lanex screen profile FWHM, of ~ 3.3 mm, corresponds to the beam width[ww].

[ww] The beam profile gets Gaussian shaped when narrow slits are used due to collimator scatter and some distance between the collimator and the scintillating GEM detector (or

Given a beam width of ~ 3.3 mm and the FWHM of ~ 3.5 mm of the scintillating GEM detector profile, the GEM detector spatial resolution is in first approximation ≤ 1 mm.

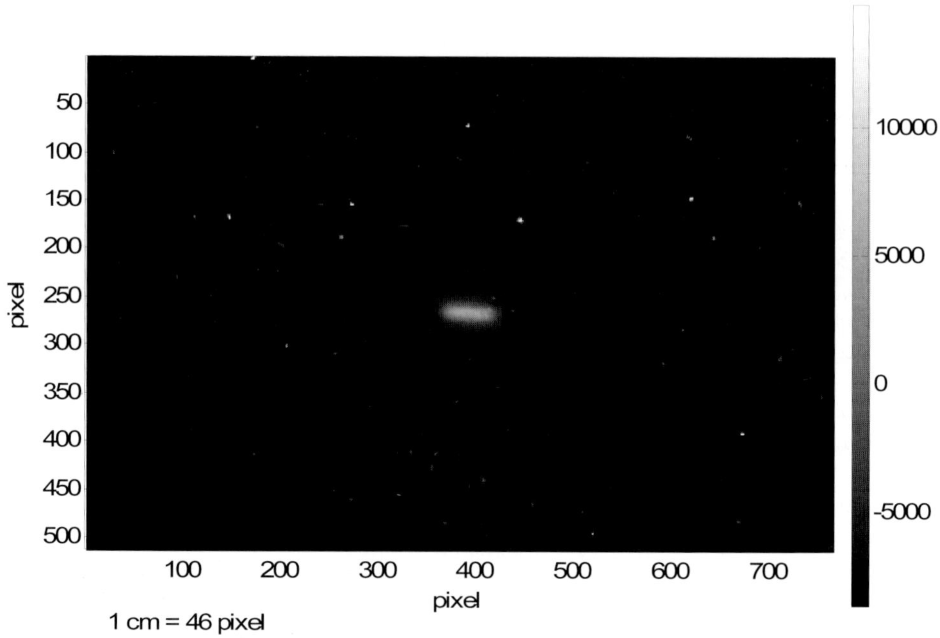

Figure 6.8 Example of scintillating GEM detector picture taken with a 10 mm long and 1 mm wide field-shaping collimator positioned almost horizontally with respect to the beam direction. GEMs with small holes were used in Ar + 8% CF_4.

The slightly wider lower part of the GEM light intensity profile is attributed to the residual background (section 4.4.2). The residual background magnitude is expected to be different for the two detectors. In the scintillating GEM detector, for example, the Al frames on which the GEMs are glued, the kapton, and the copper on the GEM surface are possible sources of light reflection, not present in the Lanex screen pictures.

Lanex). The beam width is slightly larger than the 1 mm wide slit because of some distance between the collimator and the detector.

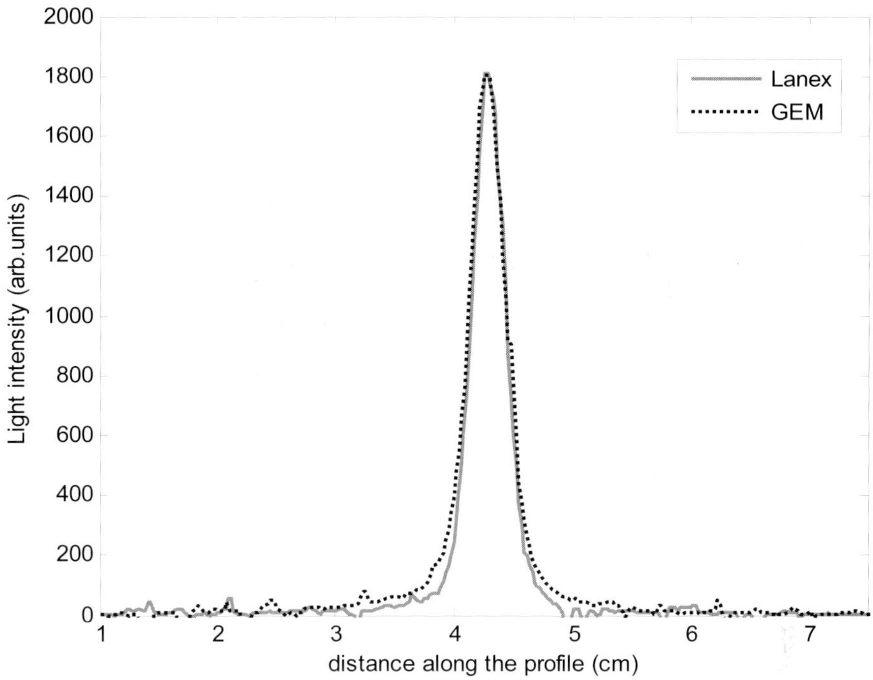

Figure 6.9 One-pixel wide vertical light intensity profile of the scintillating GEM detector picture represented in Figure 6.8, together with a Lanex light intensity profile measured under the same conditions. For visualization purposes, the latter has been adjusted (without horizontal scaling) to make the two profiles peak coincide.

6.3.6 Time response

In Figure 6.10, the rise[xx] (left graph) and fall time (right graph) of the scintillating GEM detector light signal is shown, together with the IC signal. In this case, the GEM light signal was recorded by means of a Hamamatsu R943-02 photomultiplier tube. The output of the photomultiplier tube and the IC signal were observed by means of an oscilloscope. GEMs with big holes were mounted inside the detector chamber that was flushed with Ar + 6% CF_4. The rise time of the GEM detector light signal is of the order of 2 μs. This is probably related to the switching on time of the beam [5]. In fact, the scintillating GEM detector rise time is expected to be faster since according to literature the electron drift time in 3.2 mm drift gap and in Ar + 10% CF_4 is about 0.05 μs [6]. The fall time is as fast as the rise time. Similar

[xx] The rise time is defined as the time required for the signal to increase from the 10% up to 90% of the signal height. The fall time is respectively the time needed for a signal decrease form 90 % down to 10% of the its height.

experiments were performed with the scintillating Lanex screen, and a rise and fall time of about 1 ms were measured.
The oscillations measured in the GEM detector light signal before the beam start (left graph of Figure 6.10) and just after it are probably due to the beam. To a lesser extent, they are also visible in the ionization chamber signal with a lower magnitude since this signal is much slower. The oscillations just after the beam stopped (right graph of Figure 6.10) are attributed to the electronics used to record the GEM detector light signal.

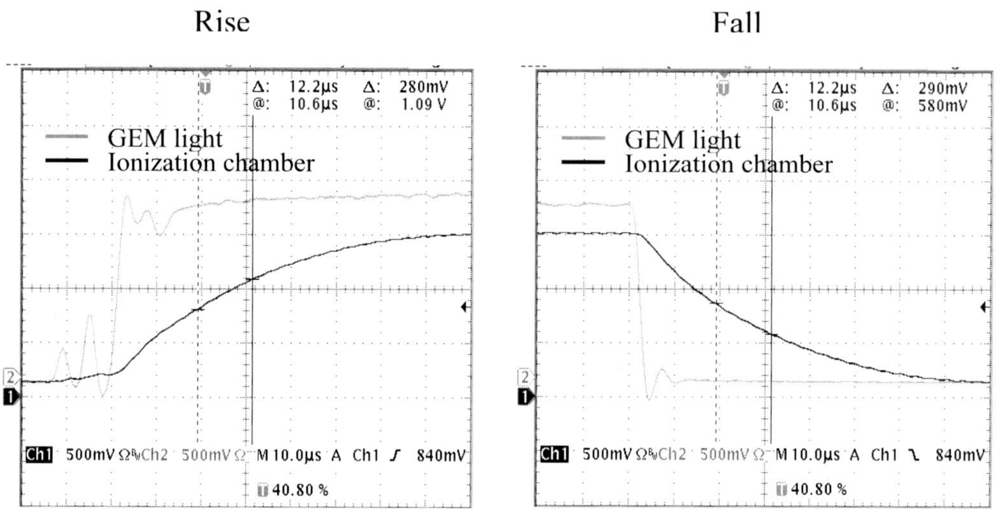

Figure 6.10 Rise time (left graph) and fall time (right graph) of the scintillating GEM detector light signal detected by means of a photomultiplier tube fed into an oscilloscope. The IC signal (dark continuous line) is graphed for comparison.

6.3.7 Pulsed beam feasibility study

In order to check if the GEM detector could work in a pulsed beam and so in a scanning beam, a feasibility study was made recording the detector outputs for a different number of beam shots. The duration and the number of the beam pulses were varied from measurement to measurement in such a way to have the same delivered dose, of about 10 Gy, per measurement and the same dose rate, of about 15 Gy/min. The CCD camera exposure time was also the same for all the pictures. Small hole GEMs were used and the detector flushed with Ar + 8 % CF_4. The field size was a circle of 20 mm^2. In Table 6.1, an overview of the results is given. L_i is normalized to IC-q in order to be sure the comparison is made between values obtained for the same delivered dose. No difference in the detector output is observed, within the experimental errors, if the 10 Gy dose is delivered in a single pulse or in a number of pulses of shorter duration and smaller Gy content. Similar behaviour of q_{out} as a function of pulse number and dose per pulse was found.

Number of pulses	Pulse duration (s)	Time between pulses (s)	Dose per pulse (Gy)	L_i / IC-q (arb.units)
1	41	-	10.25	$(1.14 \pm 0.02)\cdot 10^5$
41	1	0.1	0.25	$(1.14 \pm 0.02)\cdot 10^5$
400	0.1	0.1	0.025	$(1.14 \pm 0.02)\cdot 10^5$
4000	0.01	0.01	0.0025	$(1.17 \pm 0.02)\cdot 10^5$
40000	0.001	0.005	0.00025	$(1.17 \pm 0.02)\cdot 10^5$

Table 6.1 Feasibility study for pulsed beam application for a dose rate of about 15 Gy/min and a total delivered dose of about 10 Gy. The 1st column from the left indicates the number of beam pulses or shots delivered per measurements, the 2nd the time between pulses, and the 3rd the pulse duration. The uncertainty associated to L_i / IC-q is the repeatability error, found to be 2% in this series of measurements.

The same study of Table 6.1 was performed for a dose rate, of about 240 Gy/min, close to the one used in proton scanning beams for clinical purposes [7]. The total delivered dose per measurement was in this case of about 20 Gy. Table 6.2 summarizes the results. Also this case, all the L_i values normalized to the IC integrated signal are comparable within the uncertainties.

Number of pulses	Pulse duration (s)	Time between pulses (s)	Dose per pulse (Gy)	L_i / IC-q (arb.units)
1	5 s	-	20	$(8.59 \pm 0.17)\cdot 10^4$
50	0.1 s	0.01	0.4	$(8.68 \pm 0.17)\cdot 10^4$
500	0.01 s	0.01	0.04	$(8.82 \pm 0.18)\cdot 10^4$
5000	0.001 s	0.005	0.004	$(8.58 \pm 0.17)\cdot 10^4$

Table 6.2 Feasibility study for pulsed beam application for a dose rate of about 240 Gy/min and a total delivered dose of 20 Gy. The 1st column from the left indicates the number of beam pulses or shots delivered per measurements, the 2nd the time between pulses, the 3rd the pulse duration.. The uncertainty associated to L_i / IC-q corresponds to the repeatability error, found to be 2% in this series of measurements.

6.3.8 Energy dependence

The response of the GEM detector as a function of the energy was studied measuring a Bragg curve. The detector was equipped with small hole GEMs and flushed with Ar + 8 % CF_4. Bragg curves were measured for 31, 13 and 2 Gy/min and for circular 20 mm^2 field size. The detector response was compared to the IC and scintillating screen Bragg curves measured under the same conditions.

In Figure 6.11, an example of pictures measured at 0-wd and at the Bp-wd is shown together with the light intensity profile along a one-pixel wide horizontal line passing by the beam spot centre. The beam spot becomes wider when increasing the thickness of the water layers because of the multiple coulomb scattering (section 2.5). In both 2D pictures, the GEM edges can be seen: the vertical lines around 160 and 610 pixels, and the horizontal lines around 10 and 480 pixels.

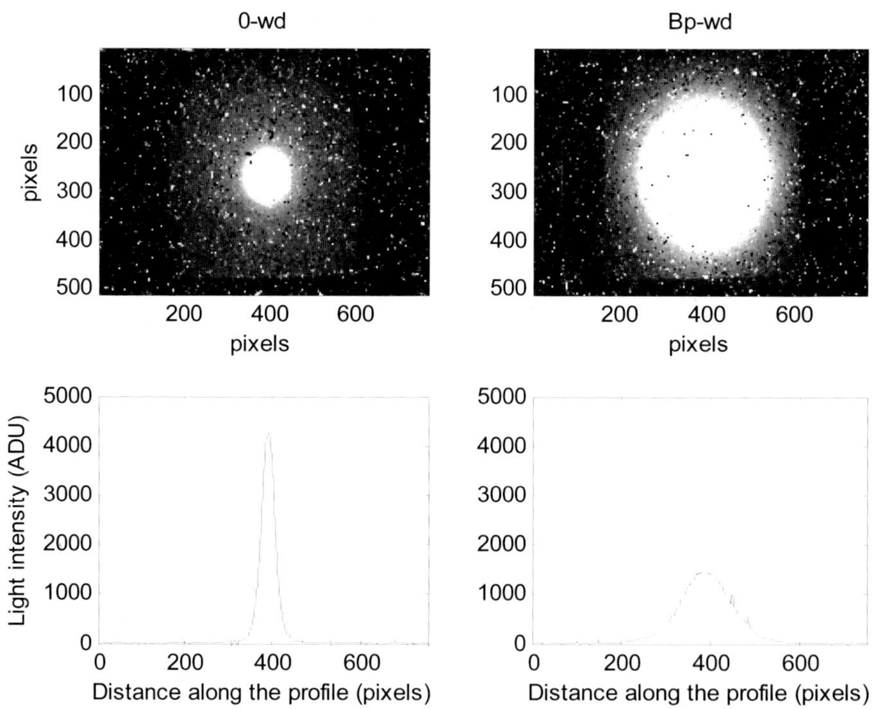

Figure 6.11 Example of pictures measured at 0-wd (on the left) and at the Bp-wd (on the right) together with the light intensity profile along a one-pixel wide horizontal line passing by the beam spot centre. These pictures were measured at 31 Gy/min. In both pictures, the vertical lines around 160 and 610 pixels and the horizontal ones around 10 and 480 pixels correspond to the GEM edges.

In Figure 6.12, the Bragg curves measured with the three detectors at 31 Gy/min are shown. L_i values were calculated for a 160 pixel radius (equivalent to 35 mm in GEM_2 facing the exit window surface plane) region of interest. Such region of interest was chosen because it corresponds to the IC sensitive area at the GEM_2 or Lanex location. All the curves are normalized to 1 at 0-wd. For visualization purposes, in the figure they are represented starting from 40 mm water depth and error bars are not shown.

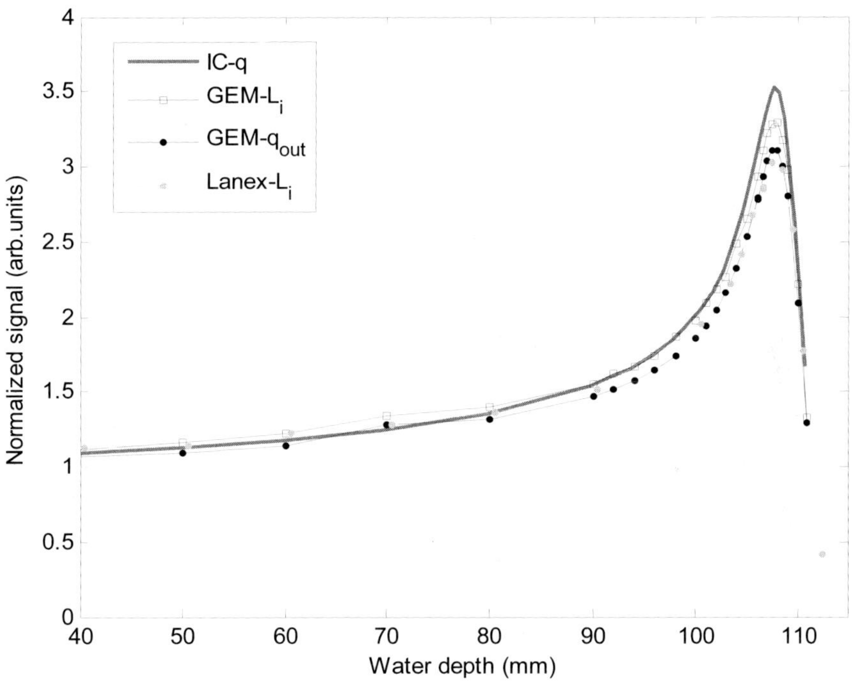

Figure 6.12 Bragg curves measured with the IC, the GEM detector and the Lanex screen. The curves are normalized to one at 0-wd. For visualization purposes, they are shown starting from 40 mm water depth.

In Table 6.3, the peak to plateau ratios of the measured curves are summarized. The IC (Figure 6.7) and Lanex [7] PPR are dose rate independent and so only the value for 31 Gy/min is reported.
The scintillating Lanex screen measured signal underestimation at Bp-wd, with respect to the IC signal, is of the same order of the one found by [7] in different beam lines.
Both L_i and q_{out} show a signal underestimation at the Bragg peak depth with respect to the IC signal. The GEM detector L_i peak to plateau ratios, measured at the three different dose rates, are compatible within the

uncertainties and the same can be concluded for the q_{out} PPR. It is difficult to quantify the signal "quenching" at the Bragg peak depth with respect to the IC signal because L_i and q_{out} Bp-wd signal underestimation is different.

	PPR
IC – q	3.53 ± 0.04
L_i – Lanex	3.03 ± 0.05
31 Gy/min	
GEM – L_i	3.32 ± 0.09
GEM – q_{out}	3.11 ± 0.09
13 Gy/min	
GEM – L_i	3.36 ± 0.09
GEM – q_{out}	3.08 ± 0.09
2Gy/min	
GEM – L_i	3.39 ± 0.10
GEM – q_{out}	3.18 ± 0.09

Table 6.3 Peak to plateau ratios measured with the IC, the Lanex screen and the scintillating GEM detector. The peak to plateau ratio of IC and scintillating screen are dose rate independent, so only one value is shown. A region of interest with 160 pixel radius was chosen for L_i calculation.

The q_{out} peak to plateau ratio is about 6 % for 31 and 2 Gy/min, and about 8 % for 13 Gy/min lower than the L_i peak to plateau ratio. The reason for this difference is not yet well understood. The signal underestimation of the detector outputs at the Bragg peak depth is expected to be the same because L_i was found to be linear with q_{out} at 0-wd as well as at Bp-wd. Furthermore, in Bragg curve measurements performed in an alpha particle beam (chapter 7) and ^{12}C ion beam (chapter 8), q_{out} and L_i PPR were found to be comparable within 1%.

L_i values were also re-calculated over a region of interest corresponding to the whole GEM surface, which corresponds to the q_{out} area of integration (or in other words area on which the electric currents flow and are measured). However, the peak to plateau ratios obtained using these "new" L_i values were not significantly different than the values reported in Table 6.3.

In Table 6.4, the q_{out} and L_i PPR of a Bragg curve measured at 13 Gy/min under the same conditions of the previous curves but without the IC in the beam, and the GEM detector at the IC position, are shown. A region of interest with 160 pixel radius was again chosen for L_i calculation. Also in this case, there is a difference between the light and electric output peak to plateau ratio, and the difference is bigger than that presented in Table 6.3. L_i PPR is within the uncertainty comparable to the peak to plateau ratio of Table 6.3 for the same dose rate. This confirms a light signal underestimation at Bp-wd of about 4-7 %. On the other hand, q_{out} PPR is not comparable and it is much lower. Peak to plateau ratios of Table 6.4 were measured after Table 6.3-data taking.

	Peak to plateau ratio
GEM - L_i	3.29 ± 0.10
GEM – q_{out}	2.74 ± 0.08

Table 6.4 q_{out} and L_i peak to plateau ratios of a Bragg curve measured at 13 Gy/min for the same conditions of the data shown in Table 6.3, but without the IC in the beam line and the scintillating GEM detector at the IC position. A region of interest with 160 pixel radius was chosen for L_i calculation.

From a careful analysis of the q_{out} values measured at 0-wd and Bp-wd without the IC in the beam, it was observed that Bp-wd q_{out} value was lower than expected. This indicates that something could have gone wrong in the electric signal measurement at this water depth, consequently influencing the PPR calculation.
Moreover, it was found that L_i PPR decreases if the region of interest radius is chosen bigger than 160 pixels. This phenomenon was not expected because the PPR should be independent of the region of interest, if the latter is chosen large enough to include the whole illuminated area. In the upper graph of Figure 6.13, L_i peak to plateau ratio is shown as a function of the region of interest radius. The peak to plateau ratio decreases by about 2.5 % when the radius is increased from 160 to 230 pixels. Although this decrement is within the peak to plateau ratio uncertainties, a comparable decrease was also observed for data measured at 31 and 2 Gy/min. To a lesser extent, this behaviour has been observed also for the Lanex screen PPR. The decrease is not related to the presence of the IC, as can be seen in the lower graph of Figure 6.13, in which PPR calculated for data measured without the IC in the beam and the GEM detector at the IC position is represented as a function of the region of interest radius. The decrease is bigger than when the IC is present. It starts at a smaller radius due to the fact that the GEM detector position was changed in between the two measurements. The corresponding Lanex screen curve without the IC in the beam line was not measured.

A possible cause of this unexpected L_i PPR decrease for "big" region of interest radii could be due to the presence of a residual background in the pictures. An attempt was made to compensate L_i values, involved in the peak to plateau ratio, for the residual background although the magnitude of the latter is not constant within a picture (section 4.4.2). A mean value of residual background was calculated using four different places in an picture for a picture at 0-wd and at Bp-wd and then subtracted from the corresponding L_i values. At the radius of 230 pixels, the correction results in a peak to plateau ratio that is about 1-1.5 % higher than the not corrected one (represented in Figure 6.13). It can be concluded that the residual background contributes to the peak to plateau ratio decrease for "big" regions of interest radii but it is not the only cause. Moreover, the reason why without the IC in the beam the decrease is bigger is not known.

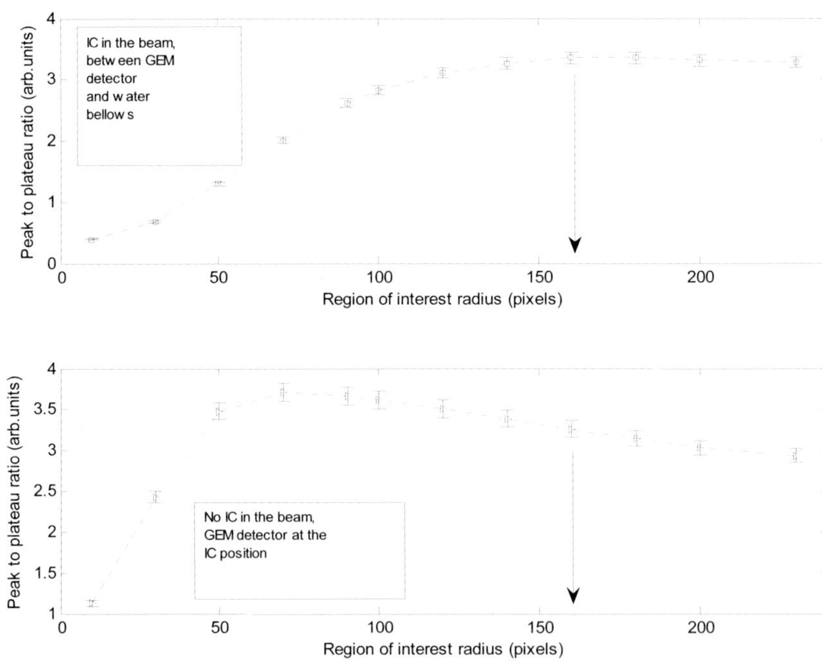

Figure 6.13 Peak to plateau ratio as a function of the region of interest radius. The data are for 13 Gy/min. Upper graph: L_i was measured with IC in the beam line. Lower graph: data collected with no IC present in the beam line and the GEM detector at IC position. The arrows indicate the radius for which data in Table 6.3 and Table 6.4 were calculated. 230 pixels radius is the biggest radius that can be taken within the GEM area.

6.3.9 Field size dependence

In the upper graph of Figure 6.14, L_i values measured with the Lanex screen and GEM detector calculated for a 250 pixel[yy] region of interest radius are shown as a function of the field size area. The field size was varied by inserting field-shaping brass collimators of different aperture size (11×12 mm^2, 20×20 mm^2, and circular ones with diameters of 30, 50, 60 and 70 mm). The experiment was performed for 5 Gy delivered dose, 6 Gy/min and at 0-wd[zz]. GEMs with small holes were mounted in the detector chamber that was flushed with Ar + 8 % CF_4. The light signal of both detectors is linear with the field size in the investigated range, as confirmed by the relative residuals of the linear least square fit represented in the lower graph of Figure 6.14.

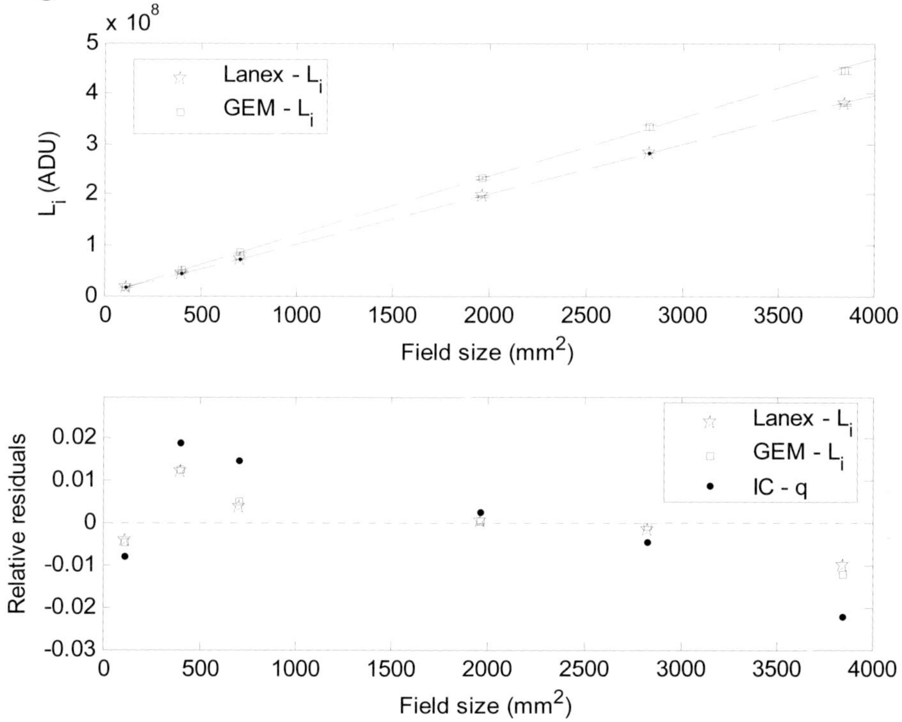

Figure 6.14 Upper graph: L_i-Lanex and L_i-GEM as a function of the field size. Lower graph: residuals of least square linear fit.

[yy] This is the biggest region of interest radius that can be taken within the picture dimensions.
[zz] A similar experiment was not performed at Bp-wd.

For a comparison, in the same plot the relative residuals of the IC integrated signal, measured simultaneously to the GEM response, are shown as well (this signal has not been represented in the upper graph of Figure 6.14 because it has a different magnitude than the L_i values). The IC-q deviation from the linear trend is similar to the one found for the GEM detector and Lanex. The higher signal for small field sizes in the relative residuals of the three detectors is probably due to collimator scattering effects.
If the GEM-L_i values are normalized to the IC integrated signal, then GEM-L_i / IC-q can be considered constant within the experimental uncertainties in the studied field size range. During the Lanex experiment, the IC signal was not recorded, so the screen L_i values can not be normalized to IC-q.

6.3.10 Response uniformity

The response uniformity was verified studying the light intensity profiles of pictures measured in a homogeneous irradiation field and using an 80 mm diameter circular collimator. An example of an picture taken under such conditions with the scintillating GEM detector, equipped with small hole GEMs and flushed with Ar + 8 % CF_4, is shown in Figure 6.15.

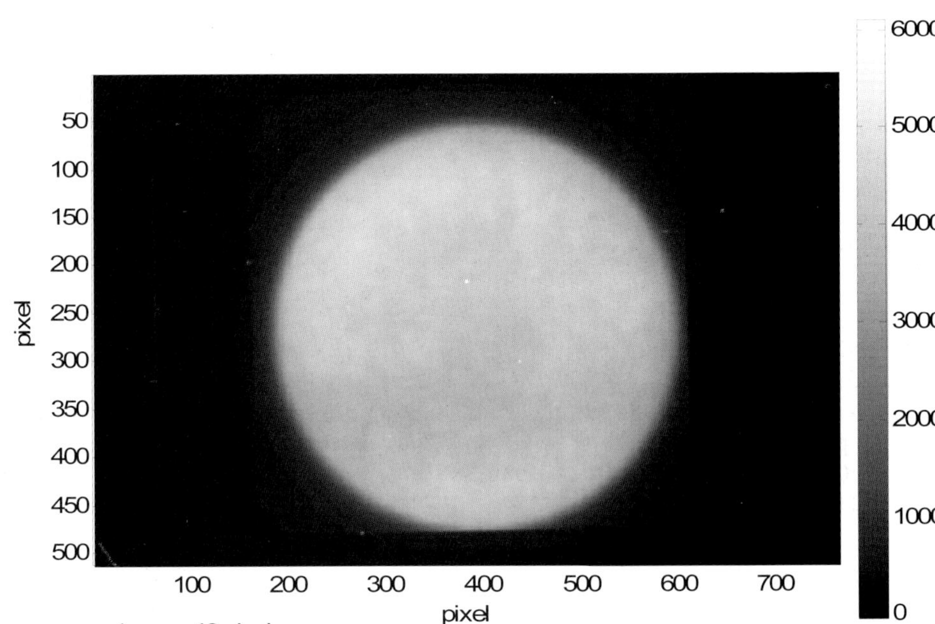

Figure 6.15 Example of picture taken with the scintillating GEM detector for an 80 mm diameter circular field-shaping brass collimator.

Some inhomogeneities in the intensity of the emitted light can be seen: below the beam spot centre and near its edges. These inhomogeneities are related to a non-complete flatness of the beam shape or to the detector response. In the upper graph of Figure 6.16, the Lanex and GEM light intensity profiles, along a one pixel wide horizontal line in the centre of the beam spot, are compared. The deviation of the GEM detector response from the Lanex screen response, along the chosen one-pixel wide horizontal line, is represented in the lower graph of Figure 6.16.

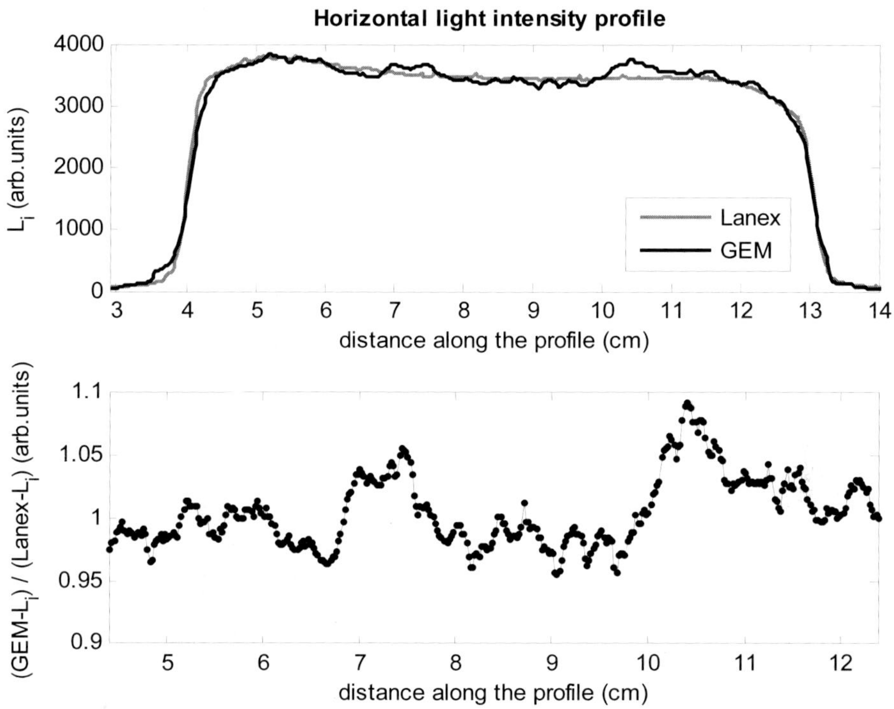

Figure 6.16 Upper graph: Lanex and GEM light intensity profiles along the same one pixel-wide horizontal line, passing by the beam spot centre, of the picture taken in the homogeneous irradiation field of an 80 mm diameter circular collimator. The GEM light intensity profile was rescaled to have the same integrated light yield as the Lanex. Lower graph: ratio of GEM and Lanex L_i along the profile. The horizontal axis is in this case restricted to the collimator size.

The Lanex signal is taken in this case as a reference; in [8], it is shown that the response uniformity of the Lanex screen is comparable to that of a film. Normalizing the GEM detector signal to that of the Lanex screen, variations in light magnitude due to the non-flatness of the beam are eliminated. Similar variations of those shown in the upper graph of Figure 6.16 were found when considering light intensity profiles along one-pixel wide horizontal lines distant ± few pixels from the line considered in Figure 6.16. Same conclusions are valid for light intensity profiles along vertical lines. Inhomogeneities in the GEM detector emitted light can be due to non uniform gain along the GEM surface [9,10] caused by, for example hole shape variations during the manufacturing process, or a non uniformly stretched GEM when it is glued onto the frame. Therefore, the uniformity of the response depends on the particual type of GEM in usage. In our case, the non uniform response, especially in the central part of the profile, can also be related to the reduced glass exit window transmission due to radiation damage (section 6.3.11), and to the variation in responsivity[aaa] of the CCD camera pixels.

6.3.11 Radiation hardness

A radiation damage effect was observed in the exit window of the GEM detector after proton beam irradiation. The glass became "dark yellow" at the beam spot position. In Figure 6.17, the transmission curve of the "dark yellow" radiation damaged part of the window together with the original glass transmission, the Ar/CF_4 emission spectrum and the CCD camera quantum efficiency are shown. The transmission of the exit window has decreased, especially in the lower part, 400 – 600 nm, of the Ar/CF_4 emission spectrum. The damage was measured after about 3600 delivered Gy in water. A correction of the L_i values for the decreased in exit window transmission is difficult to apply because the time behaviour of the darkening of the glass is not known.

6.3.12 Tissue equivalence

As discussed at point 2 of section 1.2.3, the tissue equivalence of Ar/CF_4 is better than the Lanex screen tissue equivalence but it still not ideal. Considering a mono energetic proton beam of 127 MeV initial energy, the Bp-wd signal underestimation, with respect to the "ideal" water equivalent detector, of Ar + 8% CF_4 is less than 2%, while for the Lanex screen it is about 7%.

[aaa] Due to variations in the array fabrication process, not all pixels exhibit the same sensitivity to light and a picture of a uniformly illuminated flat-field will reflect a faint checkerboard pattern at the individual pixel level. The picture processing technique of flat-fielding can be utilized to remove the pattern caused by the sensitivity variation [11].

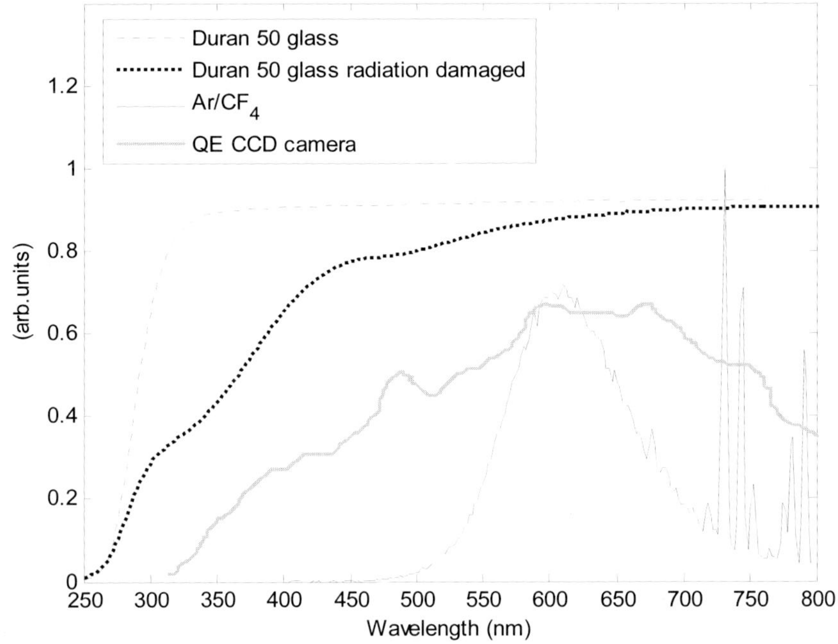

Figure 6.17 Transmission curve of the "dark yellow" radiation damaged part of the Duran 50 exit window together with the original glass transmission, the Ar/CF_4 emission spectrum and the CCD camera quantum efficiency. The Ar/CF_4 emission spectrum has been normalized to the highest spectrum value.

6.4 Conclusions

The dosimetric properties of the scintillating GEM detector were investigated in a steady proton beam. The detector was equipped either with small hole GEMs and flushed with Ar + 8% CF_4 or big hole GEMs and Ar + 6% CF_4. The response of the detector was compared to an air filled ionization chamber and/or the scintillating Lanex screen.
It was found that for small hole GEMs, flushed with Ar + 8% CF_4, and ($\Delta V_{GEM1}+\Delta V_{GEM2}$) > 716 V the integrated light signal is brighter than the Lanex screen scintillating signal. The response reproducibility of the detector output is about 4-5 % over two days. Causes of this poor reproducibility are not yet well understood and further studies should be perform on this subject especially in anticipation of a clinical usage of the detector. A good dose linearity was measured for doses between 0.05 – 19 Gy. No dose rate dependence was observed at 0-wd in the range 1 – 16 Gy/min, and at Bp-wd in the range 2 -38 Gy/min within the experimental uncertainties.

It was found that in first approximation the scintillating GEM detector has a spatial response ≤ 1 mm. The intrinsic spatial resolution of the detector is expected to be at the sub mm level because of the low Ar/CF_4 diffusion coefficient. Moreover, the combination of the high degree of granularity of the GEM holes and of the CCD camera allows imaging with high resolution. In order to measure the intrinsic GEM detector spatial resolution, an experiment should be performed using a collimator with a slit thinner than the expected spatial resolution. However, the lower part of the GEM light intensity profile is slightly wider than that of the Lanex screen. This is attributed to a residual background that could be caused by light reflections. The latter are expected to be different for the two detectors.
The GEM detector light signal has a rise and fall time, of about 2 μs, faster than the scintillating screen signal (1ms). The 2 μs rise time is probably related to the beam switching on time. Therefore, the scintillating GEM detector signal is expected to be faster. A feasibility study was performed to check if the detector could work in pulsed beam, in particular at dose rates typical of proton scanning beams. No major differences were observed in the outputs if 10 Gy at 15 Gy/min or 20 Gy at 240 Gy/min were delivered in one single pulse or different shorter pulses for the same CCD camera exposure time.
The GEM detector energy dependence was investigated by means of a Bragg curve measurement. The light signal underestimation at the Bragg peak depth is only 6 % with respect to an air filled ionization chamber and it is of the same order of magnitude of that measured for the scintillating Lanex screen (~ 14%). However, the electric output signal underestimation is about 6-8 % higher. The signal underestimation of the detector outputs at the Bragg peak depth was expected to be the same because L_i was found to be linear with q_{out} at 0-wd as well as at Bp -wd. Moreover, in Bragg curve measurements performed in an alpha particle beam (chapter 7) and ^{12}C ion beam (chapter 8), q_{out} and L_i peak to plateau were found to be comparable within 1%. From the available data, it has not been possible to find a cause for this difference. In addition, it was observed that the L_i peak to plateau ratio of the GEM detector decreases for region of interest radius bigger than 160 pixels. Although this decrement is within the peak to plateau ratio uncertainties, a comparable decrease was also observed for data measured at 31 and 2 Gy/min. Part of this peak to plateau ratio decrease is attributed to the presence of a residual background in the pictures.
No field size effects were observed in the range 120 – 3850 mm^2 within the experimental uncertainties. The non-uniform scintillating GEM detector light response over the GEM area was found to be worse than that of the

Lanex screen. Anyway, this is not a serious problem because it can be compensated taking a picture for a large field-shaping collimator every time new GEMs are mounted inside the detector chamber.
The GEM detector is in principle radiation hard but the glass exit window is not. The light transmission of the latter was found to be reduced after proton beam irradiation. A systematic study should be performed in order to quantify this radiation damage effect and it influence on L_i values. On the other hand, to eliminate this effect a quartz window can be used.
Concluding, the scintillating GEM detector looks promising as a dosimeter because it is linear with the dose, and the field size, it does not present dose rate dependence, it has a faster and brighter response than the Lanex screen and a spatial resolution of $\leq$ 0.5 mm. However, nothing definitive can be said about the proton LET dependence and further investigations are needed on this subject. Extra efforts should be made in improving the poor detector output reproducibility, and the water equivalence of the detector. The latter could be realized by means of a better tissue equivalent scintillating gas.

6.5 References

[1] http://www.kvi.nl/~agorcalc/agorhome.htm.
[2] P. van Luijk, Dose-volume effects in rat spinal cord irradiated with protons: experimental set up and mathematical modeling, PhD dissertation, University of Groningen, 2003.
[3] E. Grusell, A. Montelius, A. Brahme, G. Rikner, and K. Russell. A general solution to charged particle beam flattening using an optimized dual- scattering-foil technique, with application to proton therapy beams. *Phys. Med. Biol.*,39, 2201, 1994.
[4] P. van Luijk et al., Collimator scatter and 2D dosimetry in small proton beams, Phys. Med. Biol, 46, 2001, 653
[5] S. Brandenburg, private communication.
[6] A. Peisert and F. Sauli, Drift and Diffusion of electrons in gases a compilation, *CERN* 84-08,1984.
[7] T.Lomax, private communication.
[8] S.Boon, "Dosimetry and quality control of scanning proton beams," PhD thesis, University of Groningen, The Netherlands, 1998.
[9] R.Bouclier et al., The Gas Electron Multiplier (GEM), *IEEE Transactions on Nuclear Science*, 44, 1997, 646.
[10] B.Yu et al., Study of GEM Characteristic for Application in Micro TPC, *IEEE Transactions on Nuclear Science*, 50, 2003, 836.
[11] http://learn.hamamatsu.com/articles/ccdsnr.html.

Chapter 7

First results of a scintillating GEM detector for 2D dosimetry in an alpha beam

E. Seravalli[1], J. Hendrikse[1], J. Huizenga[2], R. Kreuger[2], J.M. Schippers[3], A. Simon[2], C.W.E. van Eijk[2]

[1] Foundation for Fundamental Research and Matter (FOM), Utrecht, the Netherlands.
[2] Delft University of Technology, Faculty of Applied Sciences, Delft, the Netherlands.
[3] Paul Scherrer Institute, Villigen, Switzerland.

IEEE, Transactions on Nuclear Science, vol. 54, 2007, 1271.

7.1 Abstract

The characterization of a scintillating GEM based gas detector for quality control of clinical radio-therapeutic beams is presented. Photons emitted by the Ar/CF_4 gas mixture are detected by means of a CCD camera; in addition, the charge is measured. The detector response has been studied as a function of alpha particle energy and dose rate. The measured signal underestimation, at the Bragg peak depth, is only few percent with respect to an air filled ionization chamber.

7.2 Introduction

The use of proton and ion therapy for tumour treatment is attracting more and more attention, due to the sharp rise and fall of the dose distribution at the end of the particle range [1]. The main issue of radiotherapy techniques is the accurate delivery of the prescribed dose to the target, while minimizing the dose in healthy tissue. In order to achieve this goal, a quality control program of beam delivery system and patient dosing is required for prevention and, in case of failure, detection of dosimetric errors [2], [3].
A relatively simple system based on a scintillating Lanex screen coupled to a CCD camera is currently used for quality control of clinical proton beams [2], [4]. The screen is mounted at the beam exit side of a phantom, whose

thickness simulates the depth in tissue. Boon [5], [6], has shown that the 2D light distribution from the screen is a valid measurement of the lateral dose distribution in the phantom at the position of the screen, in both static and dynamic beam delivery systems. However, the light emitted by the scintillating screen does not show a linear response with the dose, when the latter is changed by varying the stopping power. For instance, in a measurement of a depth dose curve, this non linear response causes a too small signal at the Bragg peak depth [5]. The signal underestimation is due to a combination of effects: the non tissue equivalent screen composition, the averaging of the signal over the finite thickness of the screen, and quenching of the light production process [7].

In an attempt to solve this problem, we are developing a gas electron multiplier (GEM) [8] based detector filled with an Ar/CF_4 scintillating mixture, and coupled to a CCD camera for dosimetry applications, as first suggested by [9]. The intensity of the measured light spot is proportional to the 2D distribution of the energy deposited in the sensitive volume by the primary electrons. GEM based scintillating gas detectors have already been used for other applications by other groups [10], [11], [12].

With a gas as primary detection medium, we expect a better linear response with the dose for the following reasons. A non tissue equivalent detector material implies that the detection-material - to - water stopping power ratio is not constant as a function of energy. In the alpha-particle energy range from 1 to 500 MeV, the Ar/CF_4-to-water stopping power ratio varies less than the scintillating screen-to-water ratio, 21 % against 55 % as can be seen in Figure 7.1. Therefore, with Ar/CF_4 the effect of the non tissue equivalence of the detection medium on the dose measurement is significantly reduced [7]. Secondly, the water equivalent thickness of a gas detector is thinner than the one of a solid state detector because of the lower density. The lower density reduces the error caused by averaging over detector thickness in strong dose gradients [7]. Finally, the light production process in a scintillating gas detector does not suffer from the quenching processes present in Lanex. In presence of incoming particles with high dE/dx, the decrease of scintillator efficiency in inorganic scintillators can be related to the depletion of available activator sites. A model for quenching is described by [13]. However, in a scintillating GEM detector photons are produced during the charge multiplication process in the GEM holes by electron-excited gas molecules which decay to the ground state [10]. Therefore, quenching processes affecting the scintillating screen response will not take place, but the light emitted by the scintillating GEM detector can be influenced by effects typical of gas proportional counters, such as gain non-uniformity, electron attachment, recombination, space charge and charge up effects. Furthermore, an afterglow (≈ 1ms) measurable in the Lanex signal is not expected in the scintillating GEM detector response due to the fast drift of the electrons in the detector (≈ 1μs) and the short life times of the excited states [14], [15].

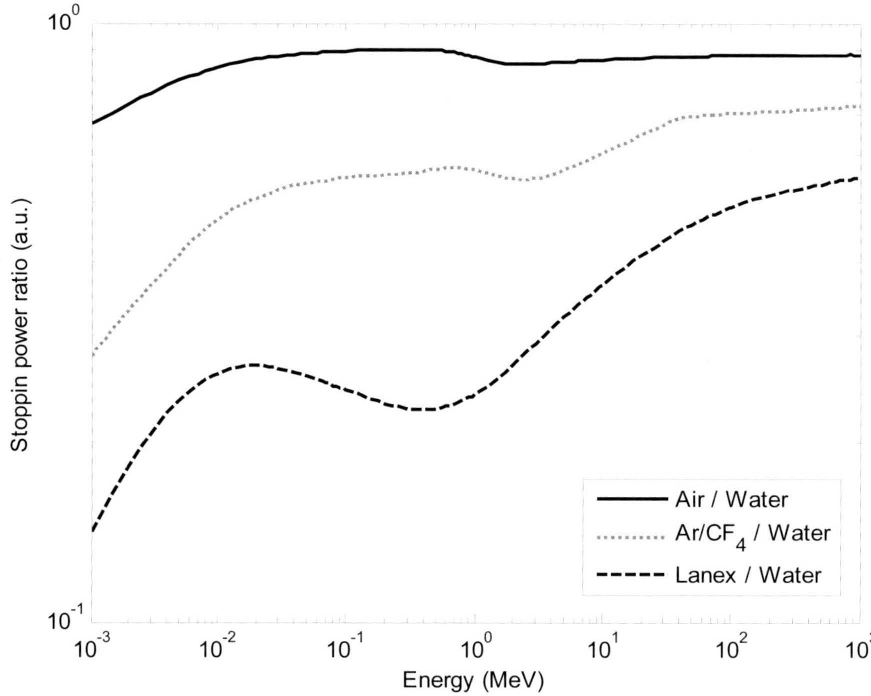

Figure 7.1 Stopping power ratios as a function of α particle energy. Solid line: air-to-water; dotted line: Ar/CF_4 (96/4)-to-water; dashed line: Lanex-to-water. Data taken from [17].

The performance in a proton beam of a detector similar to the one described here appeared already to be promising [16]. In the present work, we report on the first results of the scintillating-GEM detector operated in a 360 MeV alpha beam. The higher alpha-particle Linear Energy Transfer (LET) with respect to protons allows the study of ionization density effects on the detector response in more "extreme" conditions. In addition to that, the dose rate dependence and some scintillating GEM detector signal properties have been investigated.

7.3 Methods and Materials

7.3.1 The detector setup

The detector, represented in Figure 7.2, consists of a 6 l aluminium chamber flushed (9 lh^{-1}) with an Ar/CF_4 96/4 gas mixture at 1 atm. Inside the chamber, 2 cascaded GEMs, produced at CERN [8], [18] and named GEM_1 and GEM_2, have been mounted. The 10×10 cm^2 GEMs have 80 μm diameter double conical holes with a pitch of 140 μm and are glued onto Al frames. Aluminized Mylar foils are used as entrance window (75 μm thick) and as cathode (25 μm thick).

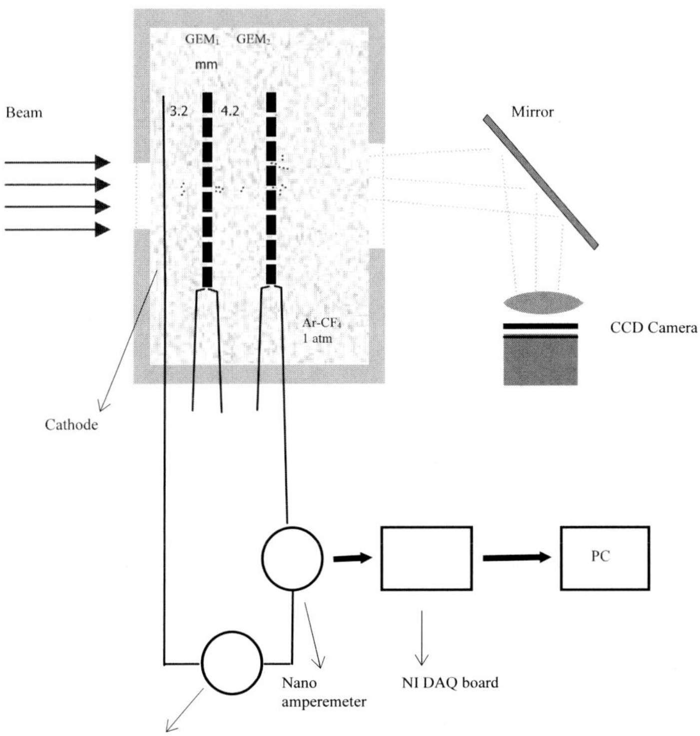

Figure 7.2 Representation, not on scale, of the scintillating GEM detector setup. For convenience, only the nano-amperemeter of the last surface of GEM_2 is represented.

The entrance window has a diameter of 70 mm. The gap between the cathode and GEM_1 (*drift gap*) is 3.2 mm, while the gap between the two GEMs (*transfer gap*) is 4.2 mm. The 100 mm diameter exit window is made

of 3 mm-thick Duran 50 glass. It is located 35 mm behind GEM_2. The cathode is grounded while each GEM surface is connected to an individual channel of a CAEN HV power supply (SY127/A231).
Nano-amperemeters, built in our electronics workshop, monitor the currents flowing to the cathode and the GEM surfaces. These meters are connected in series with the supply line of each HV channel; they have 100 kΩ impedance. A PC-controlled National Instruments DAQ board samples the measured currents every 1 ms. We define as output current, I_{out}, the current flowing to the surface of GEM_2 facing the exit window. Every experiment, apart from the ones discussed in section 7.4.1 and 7.4.2, has been performed with voltages across the GEMs, ΔV_{GEM1} and ΔV_{GEM2}, set respectively to 350V and 300V; with drift (E_d) and transfer (E_t) fields of respectively 1 kV/cm and 0.95 kV/cm. These working conditions guaranteed stable detector operation for all the used beam intensities.
Simultaneously to the electric signal monitoring, the photons produced during the electron avalanches are detected by means of a low dark-current Apogee 1E camera coupled to a Tamron 171A zoom lens. The camera has a Kodak KAF-0401E CCD with a quantum efficiency of about 62% at ~ 640 nm that matches the emission spectrum of the Ar/CF_4 gas mixture [19]. The camera is placed outside the beam to ensure low radiation background to the CCD. A 45° tilted mirror reflects the photons towards the camera. The distance between the detector exit-window and the mirror is selected such to avoid reflections from the mirror back to the window. The light-path is enclosed in a light-tight plastic tube that shields it from other light sources (e.g. LEDs in the experimental area).
The camera is focused on GEM_2 by means of a 10 cm diameter negative film, showing a 1 cm-pitch grid. The optical magnification factor of the whole set up is 0.05, given that 1 pixel ($9 \cdot 9\ \mu m^2$) on the CCD is equivalent to $176 \cdot 176\ \mu m^2$ at grid (GEM_2) position. The CCD signal per pixel is expressed in ADU (ADC unit), 1 ADU being equivalent to 9.1 electrons collected-charge (as certified by the manufacturer). During the measurements, the CCD camera is cooled down to -5°C, and the exposure time varies in the range from 40 to 60 seconds.

7.3.2 The irradiation setup

The detector has been irradiated by a 360 MeV alpha beam from the AGOR cyclotron at Kernfysisch Versneller Instituut, Groningen, the Netherlands.

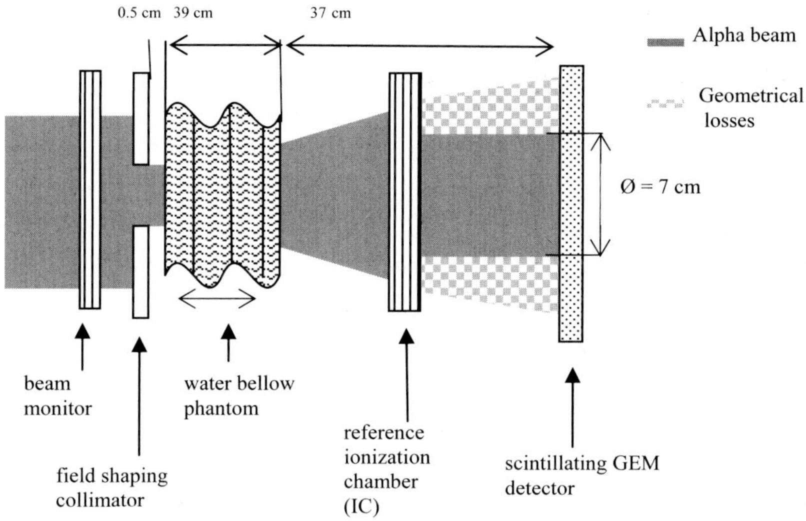

Figure 7.3 Representation of the setup for alpha irradiation, not to scale. The grey-white squared pattern area indicates the geometrical loss of alpha particles (section 7.4.3).

Figure 7.3 illustrates the irradiation setup. An air-filled parallel plate ionization chamber is used as a beam monitor to register the beam intensity (I_b) in MU/s. The MU (Monitor Unit) has been calibrated in dose to water at the scintillating GEM detector location by means of a PTW23343 Markus chamber. Assuming a 3% uncertainty in the calibration procedure, 1MU is equivalent to $(4.00 \pm 0.12) \cdot 10^{-6}$ Gy, 1MU/s = $(24 \pm 0.72) \cdot 10^{-5}$ Gy/min and 1ADU = (1.02 ± 0.03) mGy. Experiments have been done delivering doses up to 16 Gy to the sensitive volume of the scintillating GEM detector, and dose rate ranging from 10 to 240 Gy/min.
A 1.1cm × 1.2cm field–shaping brass collimator determines the beam spot at the entrance of a water-filled bellows phantom which degrades the beam energy. The bellows phantom water thickness can be varied in steps, of ≥ 0.05 mm, from zero up to beyond the alpha range. In order to check effects of ionization density on the detector response, all experiments but the one discussed in section 7.4.3 have been performed at the minimum bellows phantom water depth (0-wd), which leaves 3 cm plastic of the phantom in the beam, and at the Bragg peak water depth (Bp-wd). The charge IC-q of a second air filled parallel plate ionization chamber (IC) has been used as a reference.

7.3.3 Data acquisition and analysis

For a single measurement, the beam is turned on at a particular beam intensity I_b for a certain period of time. In that period, an amount of alpha particles, corresponding to a predetermined amount of MU seen by the beam monitor, is delivered to the detector. For each measurement, the beam monitor signal, the camera shutter and the GEM detector electric signal are sampled, as shown in Figure 7.4, and the light signal is integrated. During the experiments discussed in section 7.4.3, 7.4.4, 7.4.5 also the IC signal was monitored. The camera exposure time is chosen such to have the shutter open for a time interval slightly longer than the beam duration to guarantee the complete integration of the emitted light. The mean output current, $< I_{out} >$, is defined offline (7.1) by averaging the sampled I_{out} values between t_s and t_f, as specified in Figure 7.4. The output charge, q_{out}, is the sum of the sampled I_{out} values over the whole beam-on time (7.2).

$$< I_{out} >= \frac{\sum_{i=s}^{i=f} I_{out}(t_i)}{n} \qquad t_0 < t_s < t_f < t_e \tag{7.1}$$

$$q_{out} = \left(\sum_{i=0}^{i=e} I_{out}(t_i) \right) \cdot \Delta t \tag{7.2}$$

where t_o is the beam start time; t_e is the beam stop time; t_s and t_f are the moments in between which $< I_{out} >$ is evaluated; n is the number of samples sampled between t_s and t_f; Δt is the sample time interval. Another quantity, defined offline, is the output current normalized for the beam rate:

$$I_N = \frac{< I_{out} >}{I_b} \tag{7.3}$$

From a dimensional point of view, I_N is equivalent to a charge its units being A·s.

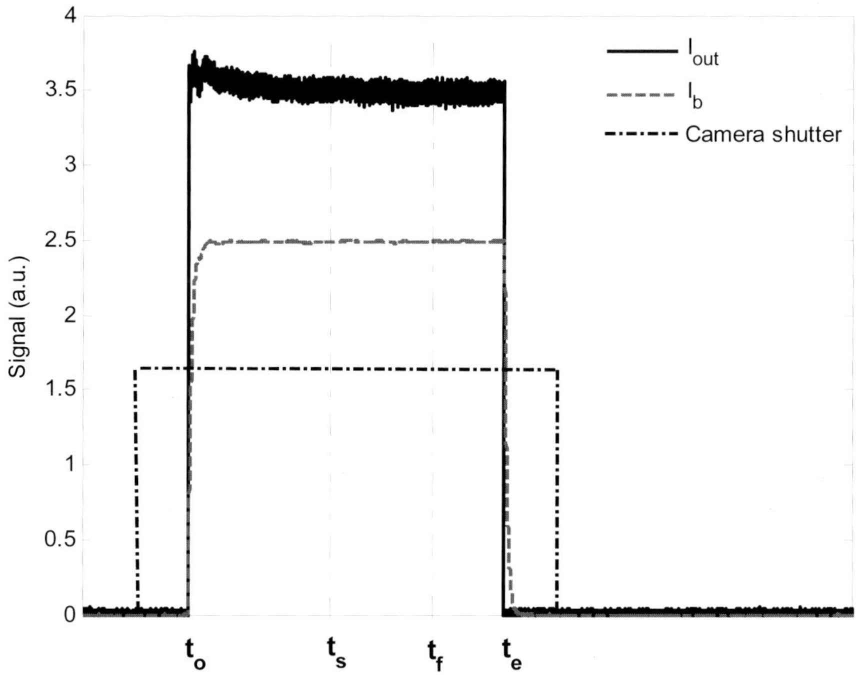

Figure 7.4 Graphical illustration of the signals recorded for each single measurement with the scintillating GEM detector. For visualization purposes of all the electric signals, only I_{out} is plotted. Solid line: I_{out}; dashed line: beam monitor; dashed-dot line: CCD camera shutter. t_o and t_e represent the beam-on and stop moments; t_s and t_f the instances in between which $< I_{out} >$ is evaluated, they define the "plateau" of the I_{out} time profile.

The CCD camera pictures are recorded and stored by means of a PC. The picture processing is made offline using Matlab routines. The CCD camera dark current noise is compensated by subtracting a picture, called background picture, taken without beam and camera shutter opened for the same exposure period. Large signals on isolated pixels are averaged out by filtering the picture with a median filter [20]. This filtering process does not influence the important features of the picture, such as amplitudes and gradients [21]. The integrated light yield values, L_i, are obtained by integrating the picture pixels over a region of interest (ROI), with a diameter equal to the entrance window diameter. In Figure 7.5, an example of a picture taken at 0-wd is shown together with the light intensity profile corresponding to the continuous black line, and the chosen ROI (dashed-dot line).

Results reported in sections 7.4.3, 7.4.4, and 7.4.5 have been corrected for the voltage drop across the resistor of the nano-amperemeters.

The uncertainty of data represents the statistical variation observed over the measured values. In the following, error bars are only shown when markers are smaller than the experimental uncertainties.

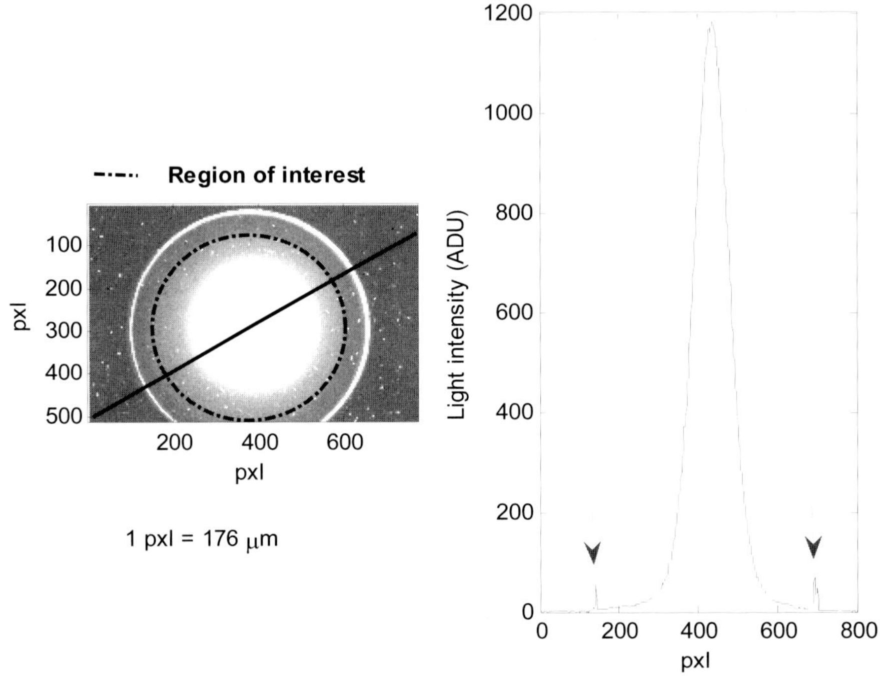

Figure 7.5 On the left, example of an picture taken with a 1.1cm × 1.2cm field–shaping brass collimator at 0-wd. The central bright circle corresponds to the beam spot, the biggest circle is light reflection on the exit window edge. The black dashed-dot line represents the ROI chosen for L_i calculation. On the right, the light intensity profile along the continuous black line is shown (the vertical arrows in the profile indicate the GEM detector exit window edge). The beam shape has become circular, due to multiple scattering in the water bellows phantom exit walls.

7.4 Results and Discussion

In section 7.4.1, and 7.4.2 scintillating GEM-based detector signal properties are discussed; in the remaining sections, dosimetric properties of the device are investigated.

7.4.1 L_i and q_{out}

In the upper part of Figure 7.6, L_i values are shown as a function of q_{out} for 0-wd and Bp-wd; q_{out} was varied by changing ΔV_{GEM1} and ΔV_{GEM2}, while keeping E_d and E_t at their default values and MU and I_b fixed. At both water depths, L_i is linearly related to q_{out}. Fraga [10] has already indicated that the observed scintillating GEM detector light is mainly due to the multiplication processes in the GEM holes, to which q_{out} is directly related. In the lower part of Figure 7.6, the residuals of least square linear fits to the 0-wd and Bp-wd data points are shown; Table 7.1 presents the fit coefficients m and z. The fits have been made for $q_{out} \leq 70$ μC for better comparison of their parameters. The bigger intercept z at 0-wd is probably due to an extra offset in the pictures, not compensated by the subtraction of the background picture. As already experienced by Simon [22], the background subtraction procedure is quite critical and can cause a systematic overestimation of L_i, especially for small signals.

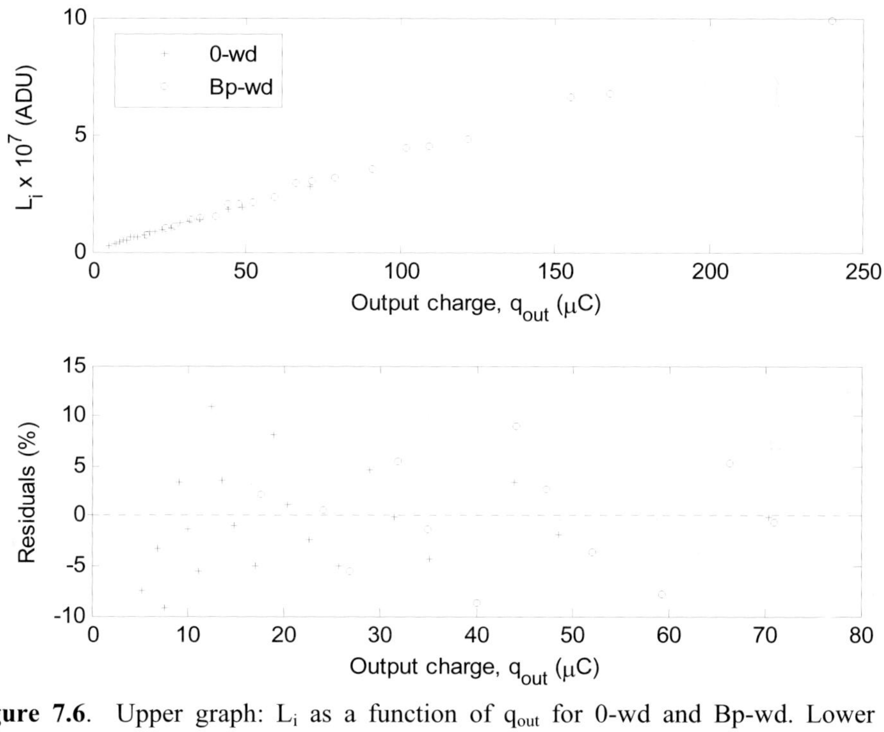

Figure 7.6. Upper graph: L_i as a function of q_{out} for 0-wd and Bp-wd. Lower graph: residuals of the two interpolating lines, calculated with a least square linear fits.

$L_i = m \cdot q_{out} + z$	0-wd	Bp-wd
$m \pm \sigma_m$	$(3.90 \pm 0.06) \cdot 10^5$	$(4.32 \pm 0.20) \cdot 10^5$
$z \pm \sigma_z$	$(7.39 \pm 1.55) \cdot 10^5$	$(-5.11 \pm 9.24) \cdot 10^5$

Table 7.1 Coefficients of least square linear fit of data of Figure 7.6 for $q_{out} \leq 70$ μC

7.4.2 Charge transfer efficiency

Table 7.2 compares q_{out} measured with larger gas amplification in GEM_1 (*1st situation*) to the values obtained with larger gas amplification in GEM_2 (*2nd situation*), for both 0-wd and Bp-wd. During the measurements, $\Delta V_{GEM1} + \Delta V_{GEM2}$ was kept at a constant value and E_d and E_t at their default values; the amount of MU and I_b were also fixed. At both depths, q_{out} is higher in the *2nd situation*. We relate this response asymmetry to the electron extraction efficiency of GEM_1, $\varepsilon_{e,1}$, defined as the fraction of GEM-multiplied electrons that are extracted from the GEM_1 holes into the subsequent gap [23], [24], in our case the transfer gap. In [23] is shown that for a fixed E_{ext}, in our case E_t, ε_e increases when decreasing the mean electric field inside the hole E_h. The latter field is linearly related to the voltage across the GEM. For example according to Figure 5 of [23], for ΔV_{GEM} = 290 V and ΔV_{GEM} = 320 V E_h is respectively of about 45 and 51 kV/cm. An identical behaviour of ε_e as a function of ΔV_{GEM} is also shown in Figure 8 of [25].

		$q_{out}(\mu C)$	
		0-wd	Bp-wd
1st situation	ΔV_{GEM1}=320V ΔV_{GEM2}=290V	(21.46 ± 0.43)	(75.34 ± 1.51)
2nd situation	ΔV_{GEM1}=290V ΔV_{GEM2}=320V	(25.93 ± 0.52)	(92.10 ± 1.84)
Ratio *2nd/1st*		1.21 ± 0.02	1.22 ± 0.02

Table 7.2 q_{out} for $\Delta V_{GEM1} > \Delta V_{GEM2}$ and $\Delta V_{GEM1} < \Delta V_{GEM2}$ measured at 0-wd and Bp-wd.

The ratio of q_{out} measured in the *1st and 2nd situation* can be described as a function of $\varepsilon_{e,1}$:

$$\frac{q'_{out}}{q''_{out}} = \frac{(q_i \cdot \varepsilon_{c,1} \cdot G_1 \cdot \varepsilon_{e,1} \cdot \varepsilon_{c,2} \cdot G_2 \cdot \varepsilon_{e,2})'}{(q_i \cdot \varepsilon_{c,1} \cdot G_1 \cdot \varepsilon_{e,1} \cdot \varepsilon_{c,2} \cdot G_2 \cdot \varepsilon_{e,2})''} \qquad (7.4)$$

where superscripts ' and " refer respectively to *1st* and *2nd situation;* while subscripts 1 and 2 refer to GEM_1 and GEM_2 respectively. q_i is the primary charge, G is the gain in a GEM, ε_c is the electron collection efficiency.

In both cases, q_i is the same since the amount of MU is fixed. In the scintillating GEM detector there is no anode, so $\varepsilon_{e,2}$ is equal to 1 according to [24]. According to [25], the product $\varepsilon_{c,1} \cdot \varepsilon_{c,2}$ can be expressed by

$$\varepsilon_{c,1} \cdot \varepsilon_{c,2} \sim e^{-x^2 \cdot E \cdot \left(\frac{1}{\Delta V^2_{GEM1}} + \frac{1}{\Delta V^2_{GEM2}}\right)} \quad (7.5)$$

where x is a constant and $E = E_t = E_d$. In the present study,

$$\left(\varepsilon_{c,1} \cdot \varepsilon_{c,2}\right)' = \left(\varepsilon_{c,1} \cdot \varepsilon_{c,2}\right)'' \quad (7.6)$$

because $E_t \approx E_d$ and $(\Delta V_{GEM1})' = (\Delta V_{GEM2})''$, $(\Delta V_{GEM2})' = (\Delta V_{GEM1})''$. Moreover, a voltage exchange for two identical GEMs gives:

$$(G_1 \cdot G_2)' = (G_1 \cdot G_2)'' \quad (7.7)$$

Thus, equation (7.4) reduces to:

$$\frac{q'_{out}}{q''_{out}} = \frac{(\varepsilon_{e,1})'}{(\varepsilon_{e,1})''} \quad (7.8)$$

In Table 7.2, q_{out}, or in other words $\varepsilon_{e,1}$, increases by ≈ 21 % when ΔV_{GEM1} is decreased from 320 V to 290 V for both 0-wd and Bp-wd. This increase is of the same order of magnitude as the one presented by the data in Figure 10 of [23] for $E_t = 0.95$ kV/cm and E_h respectively of 45 and 51 kV/cm. Thus indeed, the q_{out} difference for the two situations can be explained by the charge extraction efficiency of GEM_1. It should be noted that the measurements of [23] were made in Ar/CO_2, while the data of Table 7.2 are for Ar/CF_4.

7.4.3 Depth dose curves

In order to study the detector response as a function of the alpha energy, we measured depth dose curves, for a fixed amount of MU and at a fixed I_b. Due to energy losses in the beam line components, air, and the water phantom frame, the effective alpha-particle energy at 0-wd was 205 MeV. Additionally, we measured depth dose curves using the Lanex screen (placed at the scintillating GEM detector location) and IC. In Figure 7.7, the relative q_{out} and L_i Bragg curves are compared with the reference ionization chamber and the scintillating screen curves. The curves have been adjusted horizontally in order to have all the Bragg peak positions coinciding, and have been normalized at 0-wd.

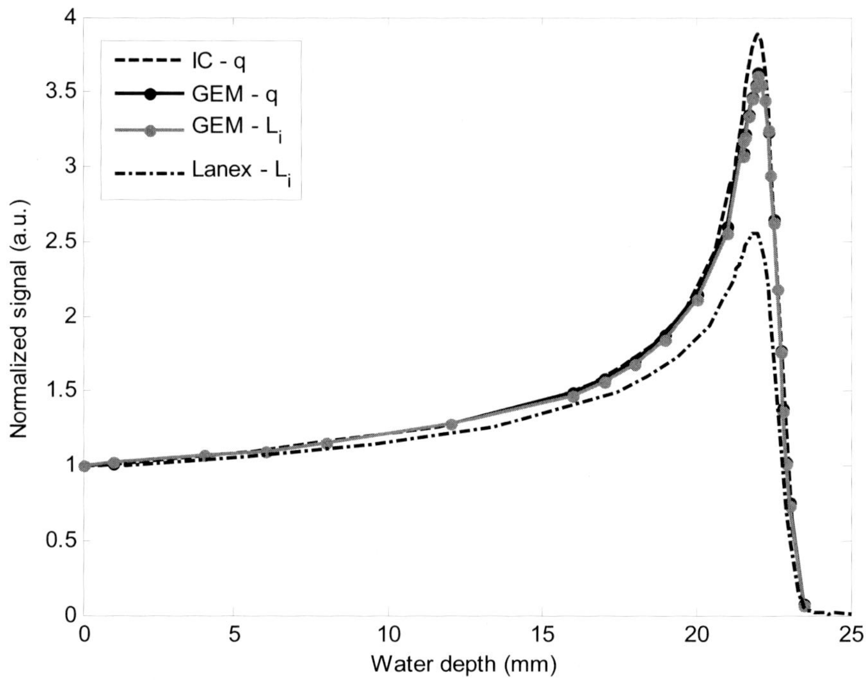

Figure. 7.7 Comparison among Bragg curves measured with the reference ionization chamber; the scintillating GEM detector and, the Lanex screen.

At the Bp-wd, the scintillating screen signal is quenched by 34 % with respect to the reference ionization chamber signal, while the underestimation of the scintillating GEM detector light signal is only 7%. The q_{out} Bragg curve is equal to the L_i curve within 1%; I_N Bragg curve was found to be equal to the q_{out} curve, therefore it is not shown in the figure. For the GEM detector, the 7% signal underestimation is partly explicable by correlating the beam spot width, which varies with the water depth, to the detector entrance window dimension. Off-line analysis has indicated that at water layers thicker than 8 mm, the beam size is larger than the entrance window. Figure 7.8 shows the beam profiles measured at the Bp-wd, with the Lanex and the scintillating GEM detector. In the profile measured with the Lanex screen there are no steps: the screen dimensions are bigger than the beam spot for all the water layer thicknesses. On the contrary, the scintillating GEM detector profile shows two sharp steps that correspond to the entrance window edges. Therefore, part of alpha particles was lost compared to the scintillating screen and the reference ionization chamber (the loss is graphically illustrated in Figure 7.3 with the grey-white square patterned area). In an attempt to evaluate the loss of signal, a Gaussian curve has been fitted to the beam profiles measured with the scintillating GEM detector at different water layers. These fitted profiles were then integrated to obtain L_i^c values, recovering in this way the area delimited by the lines at bottom left and right of the scintillating GEM beam profile of Figure 7.8.

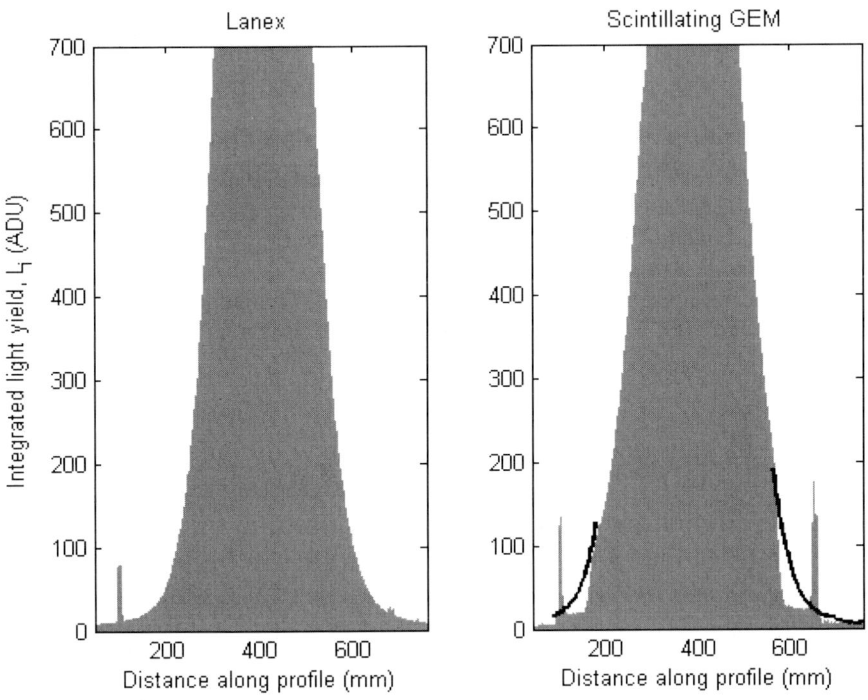

Figure 7.8 Comparison of beam profiles measured at Bp-wd with the Lanex screen and the scintillating GEM detector. The drawn lines at the bottom left and right of the scintillating GEM beam profile indicate the missing areas of the profile corresponding to the loss of alpha particles (the vertical lines in the profile indicate respectively the screen support edge and the GEM detector exit window edge).

In Figure 7.9, the Bragg peak region of the L_i^c Bragg curve is compared with the Bragg peak region of the Figure 7.7 curves. The underestimation of the scintillating GEM detector signal at the Bp-wd, including the correction, is 4 % with respect to the IC. This small signal underestimation can not be explained by the calculated 2% difference in the Ar/CF_4 – to - air stopping power ratio between 0-wd and Bp-wd, for a mono energetic alpha particle beam in the energy range of the measured Bragg curve. However, it could be related to an ionization density or dose rate dependence of the detector response.

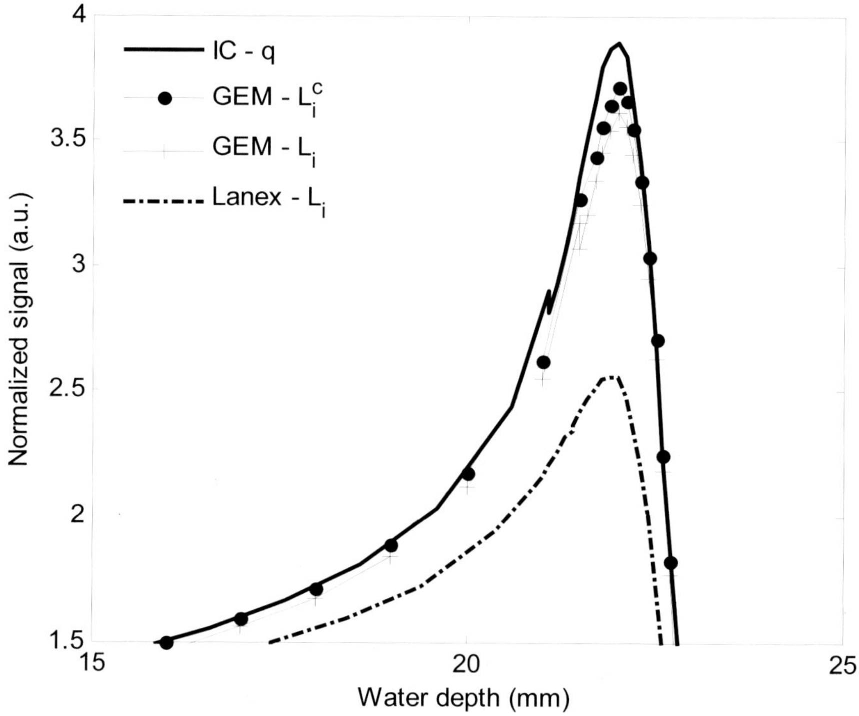

Figure 7.9 Enlargement of the Bragg peak region for the Bragg curves measured with the reference ionization chamber; the scintillating GEM detector L_i, the corrected light yield values, L_i^c, and Lanex screen.

7.4.4 Dose rate dependence

The quantities q_{out}, I_N and L_i were studied as a function of the dose rate for a fixed amount of MU. The dose rate was varied by changing I_b. Figure 7.10 presents the results measured at Bp-wd, together with the reference ionization chamber signal. The latter is constant with the dose rate as expected. q_{out} and L_i increase by 5%; on the other hand, I_N is constant with the dose rate within the uncertainties. I_N does not follow q_{out}, as it did for the Bragg curve (section 7.4.3).

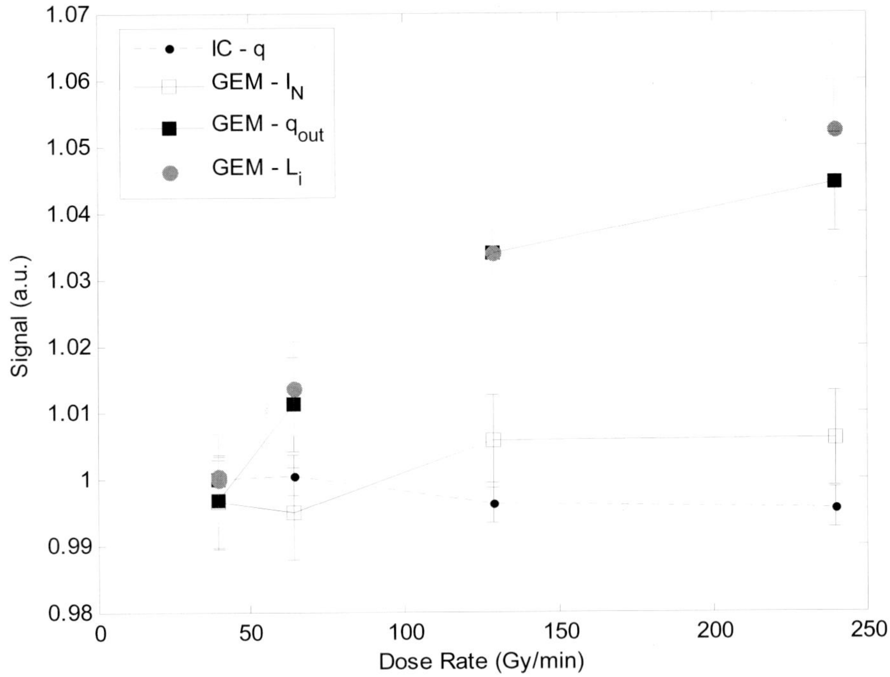

Figure 7.10 Reference ionization chamber signal, I_N, q_{out}, and L_i as a function of dose rate normalized to the respective value measured at the smallest dose rate. The data are measured at Bp-wd.

	Variation (%)	
	0-wd	Bp-wd
I_N	0.40 ± 0.56	1.11 ± 0.66
q_{out}	3.10 ± 0.97	4.55 ± 0.97
L_i	3.40 ± 0.97	4.94 ± 0.96

0-wd dose rate range: 16-62 Gy/min
Bp-wd dose rate range: 40-240 Gy/min

Table 7.3 Variation of I_N, q_{out}, and L_i among the highest and the lowest dose rate

The increase of q_{out} with the dose rate can be explained by the presence in I_{out} (Figure 7.4) of a significant overshoot in the first 1 or 2 seconds after the beam has been switched on. q_{out} is affected by it, being the sum of I_{out} between t_o and t_e. On the contrary, $< I_{out} >$, and consequently I_N are not influenced by the presence of this transient because the former is averaged between t_s and t_f. We have observed that the overshoot height, with respect

to the "plateau", increases with the dose rate; this behaviour explains the increase of q_{out} with the dose rate, and also of L_i, the two quantities being linearly related.
Also at 0-wd, q_{out} and L_i increase as a function of dose rate, while I_N remains constant within 1% (Table 7.3). This increase is again due to an initial transient in the detector response. We found only a very small difference (i.e. <0.1 %) in the overshoot height with respect to the "plateau", when the I_{out} time-profile measured at Bp-wd is compared to the 0-wd I_{out} time-profile measured for the same dose rate and appropriately normalized to the Bp-wd I_{out} "plateau". So the presence of the overshoot is not related to the alpha particles ionization density difference between 0-wd and Bp-wd. Since the reference ionization chamber signal is independent from the dose rate, we can exclude the beam being a possible cause of the initial transient. This was confirmed independently by checking the beam signal with a separate detector. In the scintillating GEM detector, several mechanisms can be responsible for the overshoot. For instance, a charging up effect could take place in the Kapton layer of the GEMs dynamically changing the electric field until a new equilibrium is reached at a reduced gain. A similar initial instability of the electrical output signal as a function of time is also shown by MSGCs [26], [27]. It is attributed to polarization effects in the substrate between the anode and the cathode. The overshoot could also possibly be caused by the electric circuit used to monitor the currents of the GEMs surfaces. With typical currents up to 34 μA, the voltages over the nano-amperometers were up to 6V.
Earlier in section 7.4.3, q_{out} and I_N Bragg curves were found to be equal: in that case, the overshoot contribution to q_{out} was negligible because the beam-on time was much longer than the overshoot duration (25 s against 2 s). The presence of an initial transient could have consequences for the application of the scintillating GEM detector in a scanning beam in which the beam pulse is shorter, e.g. 0.5 – 100 ms, than the initial transient duration.

7.4.5 Ionization density

We investigated the dependence of detector response on the way in which the energy is deposited in the drift gap for fixed dose and dose rate.
We considered the following two situations, for the same amount of MU:
1) 0-wd, high I_b.

2) Bp-wd, low I_b.

The beam intensity in the 2nd situation was chosen to have about the same dose rate of situation 1 or, in other words, to have about the same primary charge in the drift gap.
Although the same amount of energy is deposited in the detector sensitive volume, on micrometer scale the situations differ. In the 1st, the ionization density is low and the energy deposit is mainly due to the high number of alpha particles interacting with the gas. In the 2nd, the energy deposit is mainly due to the density within the tracks. Signals per second are considered in order to be independent from the integration time.
The comparison is presented in Table 7.4. $< I_{out} >$, and L_i ratios are compatible within the uncertainties. These ratios are 6-7% higher than the IC ratio, a difference that is compatible with the scintillating GEM detector uncorrected-signal underestimation measured at the Bp-wd (section 7.4.3). The scintillating GEM detector response is smaller in the 2nd situation in which the energy deposit is mainly due to the high density of the tracks. Due to the high ionization density of the tracks, recombination effects could take place decreasing the detector output.

	1st situation (high intensity)	2nd situation (high LET)	Ratio
Water Depth (mm)	0-wd	Bp-wd	-
Beam duration (s)	4	15	-
I_b (kMU/s)	249	66.5	-
IC – q (V/s)	0.366 ± 0.002	0.371 ± 0.002	0.99 ± 0.01
GEM - $< I_{out} >$ (μA)	10.82 ± 0.08 [a]	10.32 ± 0.08 [a]	1.05 ± 0.01
GEM - $L_i \cdot s^{-1}$ (ADU/s)	$(4.54 \pm 0.01) \cdot 10^6$ [a]	$(4.27 \pm 0.03) \cdot 10^6$ [a]	1.06 ± 0.01

[a] Values not corrected for the geometrical loss of signal

Table 7.4 Ionization density study

7.5 Conclusions

We have developed a scintillating gas detector using two cascaded GEMs in an Ar/CF_4 mixture. The photons emitted by the Ar/CF_4 electron-excited molecules, during the gas multiplication process, have been detected by a CCD camera. In this paper, we presented a first investigation of the properties of the scintillating GEM detector irradiated with an alpha beam.
An alpha beam has a high LET, compared to the LET of protons; consequently, it has been possible to explore the scintillating GEM detector properties in "extreme" ionization density situations.
We have found that the integrated light yield is linearly related, within the uncertainties, to the total charge extracted from the holes of 2nd GEM. The fact that for $\Delta V_{GEM1} + \Delta V_{GEM2}$ = constant, the situation with $\Delta V_{GEM1} < \Delta V_{GEM2}$ has a higher charge output can be explained relating the latter to the electron extraction efficiency of GEM_1, in agreement with the data presented in [23], [25].
After correcting for the geometrical loss of signal, at Bragg peak depth, the scintillating GEM detector signal is only 4% smaller than that of the reference ionization chamber (section 7.4.3). This result is very promising in comparison with the response of the scintillating screen, which is 34 % smaller. Consequently the scintillating GEM detector may become a feasible substitute of the Lanex screen, especially in "high L.E.T." beams like alpha or carbon ion. The small remaining signal underestimation can be explained by an ionization density dependence of the detector output, as shown in section 7.4.4 and 7.4.5. As far as the dose rate study concerns, I_N is constant with the dose rate at 0-wd and Bp-wd, whereas q_{out} and L_i increase as a function of the dose rate at both depths. Their increase is due to an initial instability in the detector response when the beam is switched on. The presence of the observed overshoot is not influenced by the different ionization densities of alpha particles at 0-wd and Bp-wd. We observed that if the beam-on time is much longer than the overshoot duration (ca. 2 s), the overshoot contribution to q_{out} and L_i is negligible. More investigations are needed in order to characterize the overshoot and understand its origin because its presence can have consequences for the application of the detector in a scanning beam in which the beam pulse is shorter than the initial transient duration.

Acknowledgments

The authors would like to thank the staff of KVI AGOR cyclotron for their support and assistance, and P. van Luijk for helping.

7.6 References

[1] A. Brahme, Recent advances in light ion radiation therapy, *Int. J. Rad. Oncol. Biol. Phys.*, 58, 2004, 603.
[2] A.J. Lomax et al., Intensity modulated proton therapy: A clinical example, *Med. Phys.*, 28, 2001, 317.
[3] E. Perdoni, The 200-MeV proton therapy project at the Paul Scherrer Institute: Conceptual design and practical realization, *Med. Phys.*, 22, 1995, 37.
[4] G.A.P. Cirrone, A fast monitoring system for radio therapeutic proton beams based on scintillating screens and a CCD camera, *IEEE Trans. Nucl. Sci.*, 51, 2004, 1402.
[5] Boon et al., Fast 2D phantom dosimetry for scanning proton beams, *Med. Phys.*, 25, 1998,464.
[6] Boon et al., Performance of a fluorescent screen and CCD camera as a two-dimensional dosimetry system for dynamic treatment techniques, *Med. Phys.,* 27, 2000, 2198.
[7] S.Boon, Dosimetry and quality control of scanning proton beams, PhD thesis, University of Groningen, The Netherlands, 1998.
[8] F. Sauli, GEM: A new concept for electron amplification in gas detectors, *Nuclear Instruments and Methods in Physics Research, A* 386, 1997, 531.
[9] J.H.Timmer et al., A scintillating GEM for 2D-dosimetry in radiation therapy, *Nuclear Instruments and Methods in Physics Research, A* 478, 2002, 98.
[10] F. A.F. Fraga et al., Imaging detectors based on the gas electron multiplier scintillation light, *Nucl. Sci. Symposium*, Conference Record IEEE, vol. 2, 1999, 829.
[11] F. A. F. Fraga, et al., Luminescence and imaging with gas electron multipliers, *Nuclear Instruments and Methods in Physics Research,* A 513, 2003, ,379.
[12] S.M.Kang et al., X-ray image acquisition based on the scintillation light from the gas electron multiplier, *Proceedings of the Eighth International*

Conference on Inorganic Scintillators and their Use in Scientific and Industrial Applications , Alushta, Crimea, Ukraine, September 2005, 292.
[13] R.B.Murray, A.Meyer, Scintillating Response of Activated Inorganic crystals to Various Charged Particles, *Phys. Rev.*, 122, 1961, 815, 1961.
[14] M.M.R.Fraga et al., Pressure Dependence of Secondary NIR Scintillation in Ar and Ar/CF_4, *IEEE Trans. Nucl. Sci.*, 48, 2001, 330.
[15] J.E.Hesser, K.Dressler, Radiative Lifetimes of Ultraviolet Emission Systems Excited in BF_3, CF_4, and SiF_4*, *The Journal of Chemical Physics*, 47, 1967, 3443.
[16] S. Fetal et al., Dose imaging in radiotherapy with an Ar–CF_4 filled scintillating GEM, *Nuclear Instruments and Methods in Physics Research, A* 513, 2003, 42.
[17] http://physics.nist.gov/PhysRefData/Star/Text/ASTAR.html
[18] http://gdd.web.cern.ch/GDD/
[19] M. M. F. R. Fraga et al., The GEM scintillation in He–CF4, Ar–CF4, Ar–TEA and Xe–TEA mixtures, *Nuclear Instruments and Methods in Physics Research, A* 504, 2003, 88.
[20] MatlabTM image processing toolbox. Mathworks Inc.,Natick MA, 1997.
[21] J.Barkhof, G. Schut, J. B. Flanz, M. Goitein, J. M. Schippers Verification of the alignment of a therapeutic radiation beam relative to its patient position, *Med.Phys.*, 26, 1999, 2429.
[22] A.Simon et al., A scintillating triple GEM beam monitor for radiation therapy, presented at the *IEEE Nuclear Science Symposium and Medical Imaging Conference*, San Juan, Puerto Rico, 2005.
[23] M. Killenberg, et al., Modelling and measurement of charge transfer in multiple GEM structures, *Nuclear Instruments and Methods in Physics Research, A* 498, 2003, 369.
[24] C. Richter, A. Breskin, R. Chechik, D. Mörmann , G. Garty, A. Sharma, On the efficient electron transfer through GEM, *Nuclear Instruments and Methods in Physics Research, A* 478, 2002, 538.
[25] G. Bencivenni, Measurement of GEM parameters with X-rays, *IEEE Trans. Nucl. Sci.*, 50, 2003, 1297.
[26] R. Bellazzini et al., The WELL detector, *Nuclear Instruments and Methods in Physics Research, A* 423, 1999, 125.
[27] J. Benlloch, et al., Further developments and beam tests of the gas electron multiplier (GEM), *Nuclear Instruments and Methods in Physics Research, A* 419, 1998, 410.

Chapter 8

A scintillating gas detector for depth dose curve measurements in clinical carbon beams

E. Seravalli[1], M. de Boer[1], F. Geurink[2], J. Huizenga[2], R. Kreuger[2], J.M. Schippers[3], C.W.E. van Eijk[2], B.Voss[4]

[1] Foundation for Fundamental Research and Matter (FOM), Utrecht, the Netherlands.
[2] Delft University of Technology, Faculty of Applied Sciences, Delft, the Netherlands.
[3] Paul Scherrer Institute, Villigen, Switzerland.
[4] GSI Gesellschaft für Schwerionenforschung mbH, Darmstadt, Germany

Physics in Medicine and Biology, to be submitted

8.1 Abstract

A two dimensional position sensitive dosimetry system based on a scintillating gas detector has been developed for pre treatment verification of dose distributions in hadron therapy. The dosimetry system consists of a chamber filled with an Ar/CF_4 scintillating gas mixture, inside which two cascaded Gas Electron Multipliers (GEMs) are mounted. A GEM is a copper clad thin kapton foil with a regular pattern of sub mm holes. The primary electrons, created in the detector's sensitive volume by the incoming beam, drift in an electric field towards the GEMs and undergo gas multiplication in the GEM holes. During this process, photons are emitted by the excited Ar/CF_4 gas molecules and detected by a mirror-lens-CCD camera system. Since the amount of emitted light is proportional to the dose deposited in the detector sensitive volume by the incoming beam, the intensity distribution of the measured light spot is proportional to the 2D hadron dose distribution. For a measurement of a depth dose curve, the scintillating gas detector is mounted at the beam exit side of a water bellows phantom, whose thickness can be varied in steps.
In this work, the energy dependence of the scintillating gas detector output signal has been verified in a 250 MeV/u clinical ^{12}C ion beam by means of a depth dose curve measurement. The measured signal underestimation at the Bragg peak depth is only 9 % with respect to an air filled ionization chamber. This is much smaller than the underestimation found for a

scintillating Gd_2O_2S:Tb ("Lanex") screen under the same measurement conditions (43%). Consequently, the scintillating gas detector is a promising device for verifying dose distributions in high LET beams, for example to check hadron therapy treatment plans consisting of different energies.

8.2 Introduction

The introduction of dynamic intensity modulation in hadron radiotherapy requires additional and efficient methods for pre treatment dose verification. On-line measurements of beam parameters, such as intensity and profile, do not suffice for the quality control of this advanced treatment technique. It is important to perform simultaneous dose measurements at many different points in the irradiation volume because the dose may be correct at some points in the field while it can be wrong at other points.
Dose measurements with a standard ionization chamber are time consuming because they require a complete application of the treatment field for each single measurement. To overcome this limitation, several methods are being developed to measure the dose in three or two dimensions. MRI gel dosimetry [1,2,3] provides 3D dose information but it has the disadvantage that a magnetic resonance imaging unit is needed for evaluation. Arrays of ionization chambers [4] present reliable dosimetric properties, but do not have satisfactory spatial resolution (~ 5-6 mm). Stacks of ionization chambers with strip-segmented anodes for 2D readout have a better spatial response but they do not provide a full 2D dose information (only two projection planes) [5,6]. The use of stacks of films [7] gives dose information with very high spatial resolution but the film measurement evaluation is time consuming. Scintillating screens [8,9] coupled to a CCD camera allow online measurements of dose distributions with a spatial resolution nearly as good as the film. However, their response suffers from saturation.
A patient treatment plan is usually composed of a series of hadron beams having different energies. This gives additional complications as, for high LET radiation, the response of gels, films and scintillating screens depends on the energy. The response of these detectors decreases for low particle energies due to saturation. As a consequence, at the end of the depth dose curve, Bragg peak depth, these detectors underestimate the dose with respect to the beginning, plateau, of the curve. A correction for this energy dependence is difficult to apply because the composition of the beam energies in the treatment plan at each position in the irradiated volume and the corresponding detector responses must be known.

We have developed a dosimetry system based on a scintillating gas detector coupled to a CCD camera for pre-treatment verification of dose distributions in hadron beams. The dosimetry system consists of a chamber filled with an Ar/CF_4 scintillating gas mixture, inside which two cascaded Gas Electron Multipliers (GEMs) [13] are mounted. A GEM is a copper clad thin kapton foil with a regular pattern of sub mm holes. The primary electrons, created in the detector's sensitive volume by the incoming beam, drift in an electric field towards the GEMs and undergo gas multiplication in the GEM holes. During this process, photons are emitted by the excited Ar/CF_4 gas molecules and detected by a mirror-lens-CCD camera system. The intensity distribution of the measured light spot is proportional to the two dimensional hadron dose distribution deposited in the detector sensitive volume. The system is a follow up of the scintillating Gd_2O_2S:Tb ("Lanex") screen setup [8,10].

With a gas as primary detection medium, we expect, in high LET radiation beams, a smaller energy dependence of the detector response compared to the scintillating screen signal. Firstly, the light production process in a scintillating gas detector does not suffer from the quenching processes present in the Lanex screen. In fact, in the scintillating gas detector the photons are emitted by electron-excited gas molecules during the gas multiplication process. Secondly, the employed Ar/CF_4 scintillating gas mixture has better tissue equivalence and a lower mass density than the scintillating screen [11]. Moreover, we expect a spatial resolution comparable to a film, and a faster and brighter response than Lanex screen.

The energy dependence of the scintillating gas detector has already been verified in a proton [12] and alpha beam [11]. In the latter beam, the measured signal underestimation, at the Bragg peak depth, was only a few percent with respect to that of an air filled ionization chamber and much smaller than the one presented by the scintillating Lanex screen. The small signal underestimation was attributed to the ionization density of the alpha particle tracks [11].

In the present work, we report on the first results of the scintillating gas detector operated in a clinical ^{12}C ion beam and we compare its response along a depth dose curve to that of an air filled ionization chamber and to that of a scintillating Lanex screen.

8.3 Methods and Materials

8.3.1 Detector setup

The detector, represented in Figure 8.1, consists of a 350×350×50 mm^3 aluminium chamber continuously flushed (9 $l \cdot h^{-1}$) with an Ar/CF_4 92/08 gas mixture at 1 atm. Inside the chamber, two cascaded Gas Electron Multipliers (GEMs), produced at CERN [13], and named GEM_1 and GEM_2, have been mounted. The 100×100 mm^2 GEMs with 60 μm diameter double conical holes at a pitch of 90 μm, are glued onto Al frames. 25 μm thick aluminized Mylar foils are used as entrance window and as cathode. The entrance window is 150×150 mm^2. The gap between the cathode and GEM_1 (*drift gap*) is 3.2 mm, while the gap between the two GEMs (*transfer gap*) is 4.2 mm.

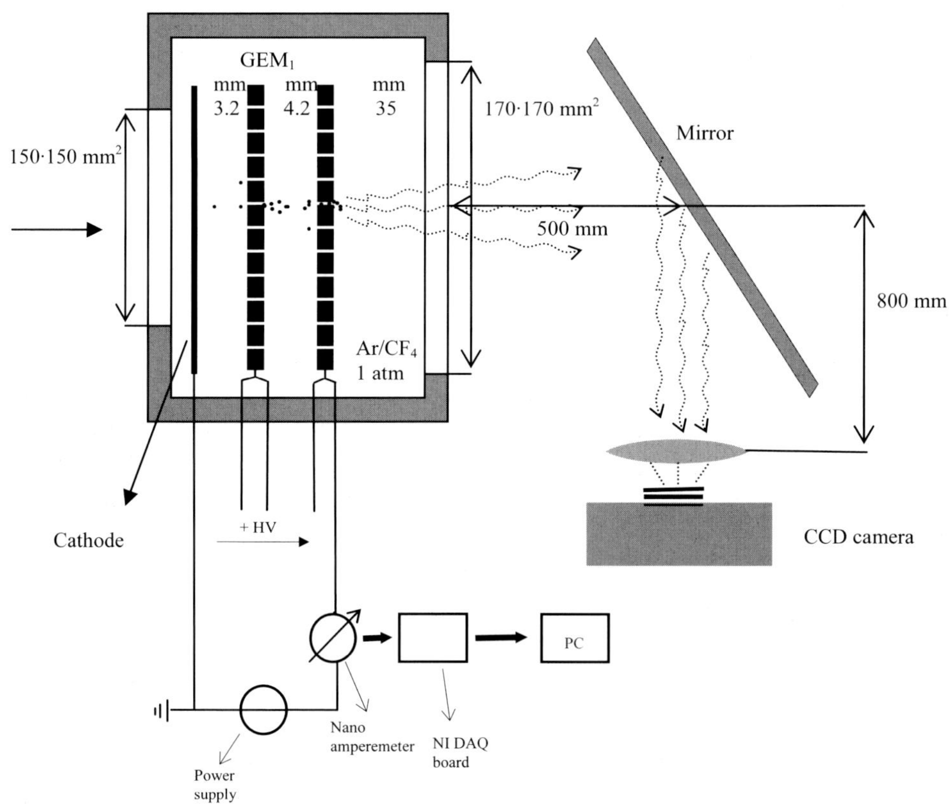

Figure 8.1 Representation, not to scale, of the scintillating GEM detector setup. For visualization purposes, only the nano-amperemeter of the last surface of GEM_2 is shown.

The 170×170 mm^2 exit window is made of 3 mm-thick Duran 50 glass and it is located 35 mm behind GEM_2. The gap between GEM_2 and the exit window is named *light gap*.
The photons produced during the electron avalanches are detected by means of a low dark-current Apogee 1E camera coupled to a Tamron 171A zoom lens. The camera has a Kodak KAF-0401E CCD with a quantum efficiency of about 62% at ~ 640 nm that matches the emission spectrum of the Ar/CF_4 gas mixture [14]. The CCD camera pictures are recorded and stored by means of a PC. The camera is placed outside the beam to ensure low background radiation onto it. A 45° tilted mirror reflects the photons towards the camera. The distance between the detector exit-window and the mirror is chosen such to avoid reflections from the mirror back to the window. The light-path is enclosed in a light-tight plastic tube that shields it from other light sources in a treatment room. The CCD camera is focused on GEM_2 by means of a 10 cm diameter transparent foil with a 1cm pitch-grid, temporarily mounted at the GEM_2 location. The optical magnification factor of the whole set up is 0.043, giving that 1 pixel (9 μm × 9 μm) on the CCD is equivalent to 207 μm × 207 μm at the transparent foil position. The CCD signal per pixel is expressed in analog to digital units (ADU), 1 ADU being equivalent to 8.4 electrons collected-charge on the CCD camera. During the measurements, the CCD camera is cooled down to – 20 ° C.
Simultaneously to the light signal, the cathode and GEMs currents are measured for a better understanding of the detector operation. The cathode is grounded while each GEM surface is connected to an individual channel of a positive CAEN HV power supply (SY127/A231). Nano-amperemeters, built in our electronics workshop, measure the currents flowing to the cathode and the GEM surfaces. These meters are connected in series with the supply line of each HV channel. The nano-amperemeter monitoring the current flowing on the cathode has 10 MΩ impedance; the ones measuring the currents flowing on GEM_1 and GEM_2 surfaces have respectively 200 kΩ and 56 kΩ impedance (that gives a precision of tens of nA on the current measurement). These impedance values were chosen to have a well detectable signal and at the same time a negligible voltage drop across them (< 0.5 V) for the expected beam fluxes. A PC-controlled National Instruments DAQ board samples the measured currents every 1 ms.
All experiments have been performed with voltages across the GEMs, ΔV_{GEM1} and ΔV_{GEM2}, set respectively to 350V and 340V; with drift (E_d) and transfer (E_t) fields of respectively 1 kV/cm and 1.5 kV/cm. These working conditions guaranteed stable detector operation for the used beam rate.

The temperature and pressure inside the detector were monitored by means of a sensor mounted on the detector chamber.

8.3.2 Irradiation setup

The experiments were performed in the clinical beam line (cave-M) for carbon ion radiotherapy treatment at the synchrotron of GSI, Darmstadt, Germany [15]. In Figure 8.2, the irradiation setup is shown.

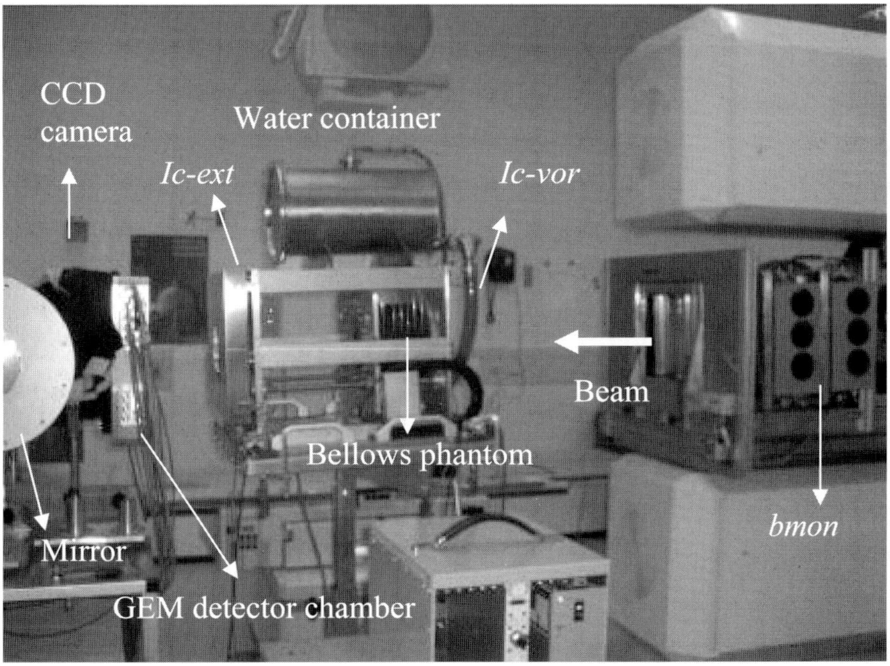

Figure 8.2 Scintillating GEM detector irradiation setup in cave-M of Gesellschaft für Schwerionenforschung mbH, Darmstadt, Germany, used for clinical carbon ion radiotherapy treatment. The patient table has been moved to the side and the GEM detector positioned at the beam line isocentre.

The scintillating GEM detector was set at the isocentre of the beam line. Between the nozzle and the GEM detector, a phantom made by a water column with remotely controlled variable thickness (0.1 mm precision), sandwiched between two air filled parallel plate ionization chambers (*Ic-vor* and *Ic-ext* with respect to the beam direction) was positioned to simulate the depth in tissue. The GEM detector was positioned horizontally and vertically such to have the beam spot in the detector centre. The signal of the *Ic-ext* was used as a reference, and it was recorded by means of a

Keithley charge meter during each spill and then summed over the number of spills during a measurement.
The scintillating GEM detector was irradiated with a 250 MeV/u ^{12}C beam with a ripple filter in the beam path, i.e. a condition used in clinical practice. The insertion of the ripple filter in the beam causes an increase of the beam energy spread as well as a small decrease of the average beam energy, in order to get the sharp carbon ion Bragg peak broader and lower [16].
The beam was pulsed: the duration of one pulse (spill) was about 4s with a 50 % duty cycle (~ 2s beam on). Experiments were performed at a rate, used in clinical practice, of $2 \cdot 10^6$ ions/spill.
The detector light and electric outputs were integrated over 20 spills which correspond to about 1 Gy totally delivered dose in water. The dose per spill was calculated multiplying the number of ions per spill by the ion stopping power and dividing by the beam spot area. The FHWM of the beam spot at the minimum water phantom thickness, *plateau*, was about 1 cm, and at the Bragg peak depth about 2.5 cm.
The CCD exposure time, of 96 s, is set to be slightly longer than the beam duration in order to guarantee the complete integration of the emitted light.
The signal of an air filled ionization chamber (*patient monitor*), positioned in the cave beam nozzle, was used as a beam monitor, *bmon*. The pulses of the *patient monitor* were directly counted by a counter on the PC-controlled National Instruments DAQ board which is sampled every 1 s.
During the measurements with a standard film and a Lanex scintillating screen, these detectors were placed at the GEM_2 location with respect to the beam. The Lanex screen Bragg curve shown in Figure 8.7 and Figure 8.8 was measured at a different moment than the other curves, and with a different camera [17] but for identical beam characteristics, as described above.

8.3.3 Data acquisition and analysis

For a single measurement, the beam is turned on and 20 spills are delivered. As can be seen in Figure 8.3, during this time *bmon*, and the GEM detector currents are sampled on the National Instruments DAQ board. The emitted light is integrated on the CCD for the exposure time set. The obtained pictures are processed offline using Matlab routines. The CCD camera dark current, the light produced in the glass exit window, and Ar/CF_4 scintillation in absence of gas multiplication are compensated for by subtracting a so called *background picture*. This *background picture* is taken with beam on, camera shutter open for the same exposure period as for a normal picture, and ΔV_{GEM1} and ΔV_{GEM2} set to zero volt while E_d, E_t and the *light gap* electric field set to the standard values.

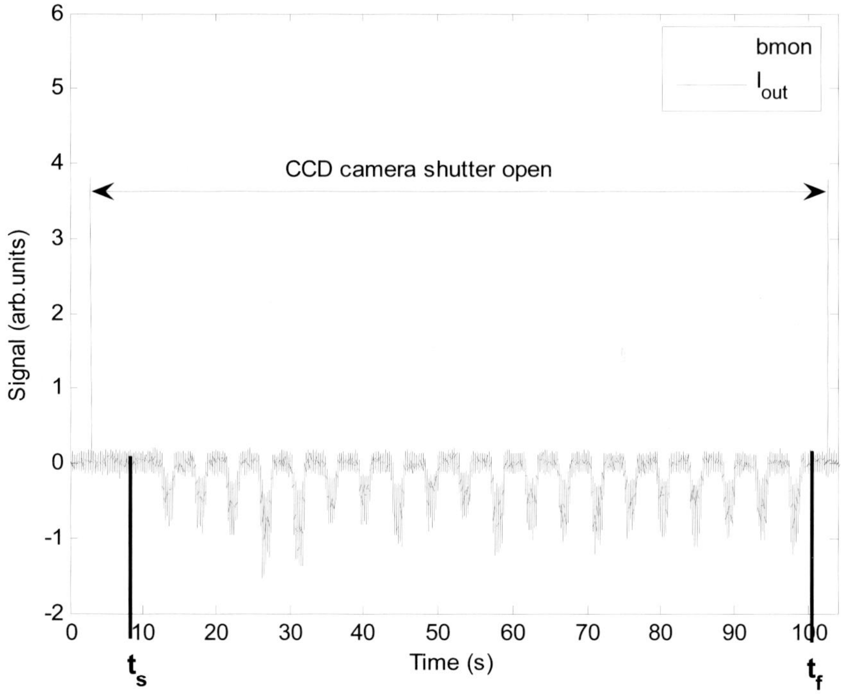

Figure 8.3 Example of the signals recorded for each single measurement with the scintillating GEM detector. For visualization purposes of all the electric signals, only I_{out} is represented. CCD camera shutter open is indicated. Dashed line: *bmon* (positive signal); solid line: I_{out} (negative signal). t_s and t_f are the instances in between which q_{out} is evaluated.

A 3-by-3 median filter [18] is applied to the pictures after background subtraction in order to remove large signals on isolated pixels created by the direct interaction of scattered radiation in the CCD. The integrated light yield, L_i, has been calculated by integrating the background corrected picture pixel values in ADU over a circular region of interest. The region of interest is chosen bigger than the beam spot recorded in a picture taken at the Bragg peak depth and it is kept constant for all the pictures.
In Figure 8.4, an example of a picture measured at the minimum water phantom thickness, is shown together with the light intensity profile along the one pixel-wide dashed white line. The continuous black line represents the region of interest over which the pixel values were integrated.

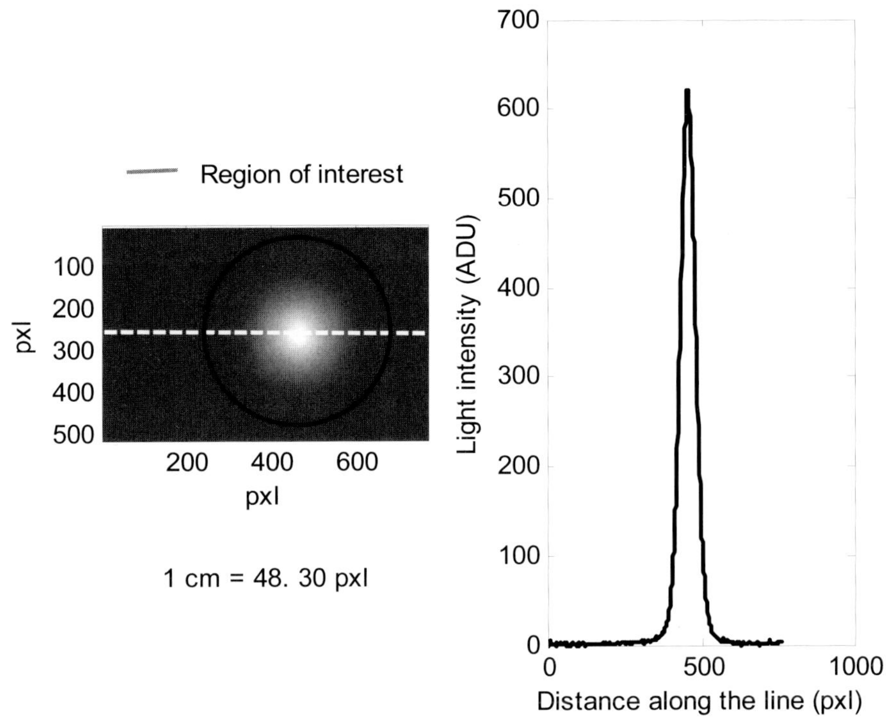

Figure 8.4 On the left, example of picture measured in carbon ion beam at the minimum water phantom thickness. The continuous black line represents the region of interest over which the pixels are integrated. On the right, the light intensity profile along the one pixel-wide dashed white line is shown.

We define as output current, I_{out}, the current flowing to the surface of GEM_2 facing the exit window. The I_{out} offset, $< I_{offset} >$, is calculated taking the mean value over N_1=2000 samples of $I_{out}(t_i)$ recorded before the beam starts.

$$< I_{offset} > = \frac{\sum_{i=1}^{i=N_1} I_{out}(t_i)}{N_1} \quad \text{with} \quad t_{N_1} < t_s \tag{8.1}$$

The output charge q_{out} is evaluated summing the offset corrected I_{out} values between t_s (beam starts) and t_f (beam stops) instants, as defined in Figure 8.3. Between t_s and t_f instants, N_2 samples are measured. Δt is the sampling time, 1ms.

$$q_{out} = \left(\sum_{i=s}^{i=f} \left(I_{out}(t_i) - < I_{offset} > \right) \right) \cdot \Delta t \quad \text{with } t_s < t_f \tag{8.2}$$

For a depth dose curve measurement, data were taken for increasing water phantom thickness, starting at the minimum phantom thickness of 57 mm water equivalent[bbb] up to the carbon ions range.
In the following, in order to compensate for beam intensity fluctuations L_i, q_{out}, and *Ic-ext* of each measurement are normalized to the *Ic-vor* integrated signal. It has been verified that normalizing L_i, q_{out}, and *Ic-ext* to the *bmon* integrated signal leads to the same relative results. For all the detectors used, the *peak to plateau ratio* is defined as the ratio of the signal measured at the Bragg peak depth and the one measured at the minimum phantom thickness, *plateau* of the depth dose curve.

8.3.4 The light signal

The recorded light signal S per carbon ion on the CCD camera is proportional to the Ar/CF_4 photon yield, Y. According to literature [19], the latter is expressed in *number of photons per secondary electron. Secondary electron* indicates, in this case, the total number of electrons after gas multiplication by the two cascaded GEMs which is collected on the GEM_2 surface facing the exit window. In other words, it is the product of the number of primary electrons created per carbon ion in the *drift gap* (n_e ~ 2092 at the *plateau*) and the detector gas gain, G. The relationship between S and the Ar/CF_4 photon yield can be described as:

$$S = f \cdot Y \cdot G \cdot n_e \qquad \text{(ADU/carbon ion)} \tag{8.3}$$

where f is the proportionality factor given by the product of the inverse of the CCD camera gain, 0.12 ADU/e^-, and the probability that a photon reaches the CCD camera and interacts with its pixels creating an electron. This probability, ~ $2.7 \cdot 10^{-6}$, takes into account the CCD camera quantum efficiency, the transmission of the optical elements and the optical solid

---

[bbb] The peak to plateau ratio value depends on the water depth at which the depth dose curve is normalized to one. By normalizing the depth dose curve at 57 mm water equivalent depth instead of 0 mm water depth, an "error" of about 8% is made on the peak to plateau calculation.

angle.
Given the spread of Ar/CF_4 photon yield values reported in literature [12,14], it is difficult to make an *a priori* estimation of S. Moreover, the gas gain G is not easy to measure because it depends on the Ar/CF_4 ratio, the gas mixture purity, the type of GEMs employed, and ambient conditions.

8.3.5 Uncertainties

The L_i statistical error is calculated taking into account the noise associated with the picture acquisition by means of the CCD camera. The main noise sources are [20,21]:

- The noise associated with the random arrival of photons at any detector. It is the square root of the number of collected photons, since the arrival of photons is governed by Poisson statistics.
- The dark current noise. Although the dark current signal can be corrected for, the noise associated with this signal can not. The dark current noise is equal to the square root of the dark current signal.
- Readout noise, which is noise produced by the on-chip amplifier and other sources of noise in the data transmission before the signal is converted into a digital representation by the ADC in the CCD.

The q_{out} statistical error is evaluated by means of the error propagation formula:

$$\sigma_{q_{out}} = \Delta t \cdot \sqrt{N_2 \cdot \sigma^2_{I_i(on)} + N_2^2 \sigma^2_{<I_{offset}>}} \tag{8.4}$$

where N_2 is the number of samples between t_s and t_f, $\sigma_{I_i(on)}$ is standard deviation of the sampled values $I_{out}(t_i)$ when the beam is on and $\sigma_{<I_{offset}>}$ is the error of $< I_{offset} >$. The latter is calculated according to the following equation:

$$\sigma_{<I_{offset}>} = \frac{1}{\sqrt{N_1}} \sigma_{I_i(off)} \tag{8.5}$$

where $\sigma_{I_i(off)}$ is the standard deviation of the samples on which the offset is evaluated by means of formula 8.1.

A distinction is made between $\sigma_{I_i(off)}$ and $\sigma_{I_i(on)}$ because it was found that the noise in the sampled values is larger when the beam is on, probably due to interference picked up from the accelerator system.
The uncertainty on the *peak to plateau ratio* is calculated by means of the error propagation formula, taking into account that *Ic-vor* integrated signal has a precision of 1.5 %, while *Ic-ext* signal 2.5 % [22]. The *peak to plateau ratio* uncertainty is mainly due to the *Ic-vor* precision.

8.4 Results and Discussion

8.4.1 Scintillating GEM detector response in a pulsed beam

In Figure 8.5, the q_{out} and *Ic-vor* integrated signals are represented as a function of the spill number for a measurement performed at the Bragg peak depth. In this particular case, q_{out} and the integral of *Ic-vor* signal are calculated for each spill, i.e. every ~ 4 seconds.
q_{out} scales with *Ic-vor* signal, or in other words it follows very well the beam intensity fluctuations. The same study could not be performed on L_i values, because the emitted light is integrated over 20 spills. Anyway, L_i is expected to have the same behaviour as q_{out} since, as shown in [11], it is linearly related to the latter.

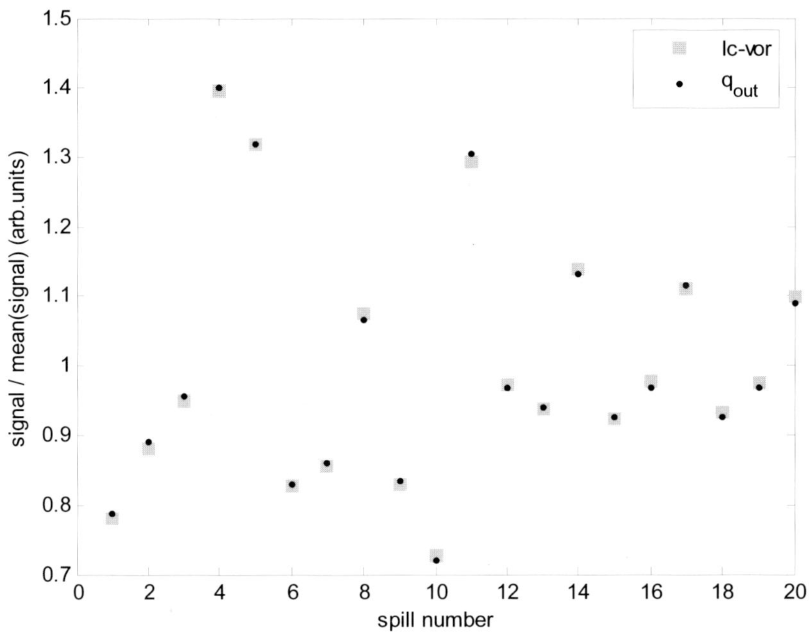

Figure 8.5 q_{out} and *Ic-vor* calculated for each spill of a Bragg peak depth measurement consisting of 20 spills.

8.4.2 Light intensity and spatial response

At the *plateau*, the scintillating GEM detector light intensity was found to be about 3 times higher than the Lanex screen signal for identical measurement conditions. Part of this higher signal can be attributed to the fact that the screen response is already "quenched" at the minimum water phantom thickness in a ^{12}C ion beam [17].

In Figure 8.6, the light intensity profile of a picture taken with the scintillating GEM detector at the *plateau* is compared to a standard film optical density profile. The two profiles were taken under the same measurement conditions, and they have been adjusted (without horizontal scaling) in order to make their peaks coinciding. The two profiles show the same FWHM. The beam FWHM, measured by the film, is about 10.4 mm. The FWHM of the scintillating GEM detector profile is about 10.9 mm. Consequently, the scintillating GEM detector has in first approximation a spatial resolution that is ≤ 3.3 mm. However, the full width at one tenth of the maximum (FWTM) is 3 mm wider than that of the film profile. This indicates the presence of tails in the spatial distribution of the scintillating GEM detector.

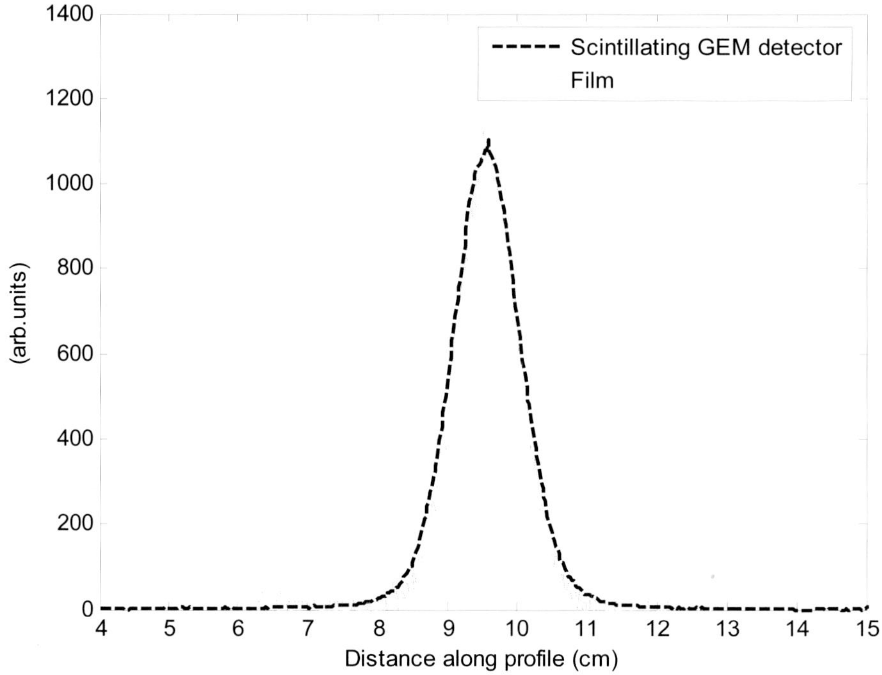

Figure 8.6 Light intensity profile of a picture taken with the scintillating GEM detector at the *plateau* compared to a standard film optical density profile, measured under identical conditions. For visualization purposes, the profiles have been horizontally shifted and vertically normalized in magnitude in order to make the two peaks coincide.

8.4.3 Depth dose curve

In order to study the scintillating GEM detector energy response, we measured a depth dose curve. In Figure 8.7, the relative L_i and q_{out} Bragg curves are compared to the *Ic-ext* curve measured simultaneously, and to the scintillating screen curve measured in a different moment [17] for the same beam conditions.

The curves have been adjusted horizontally in order to have all the Bragg peak positions coinciding, and have been normalized to one at the minimum water depth thickness of 57 mm. An enlargement of the Bragg peak region of the curves represented in Figure 8.7 is shown in Figure 8.8. The q_{out} Bragg curve is equal to the L_i curve within the uncertainties, like it was found in [11]. As can be seen in Table 8.1, the scintillating GEM detector signal underestimation at the Bragg peak depth is about 9 % with respect to the *Ic-ext* signal.

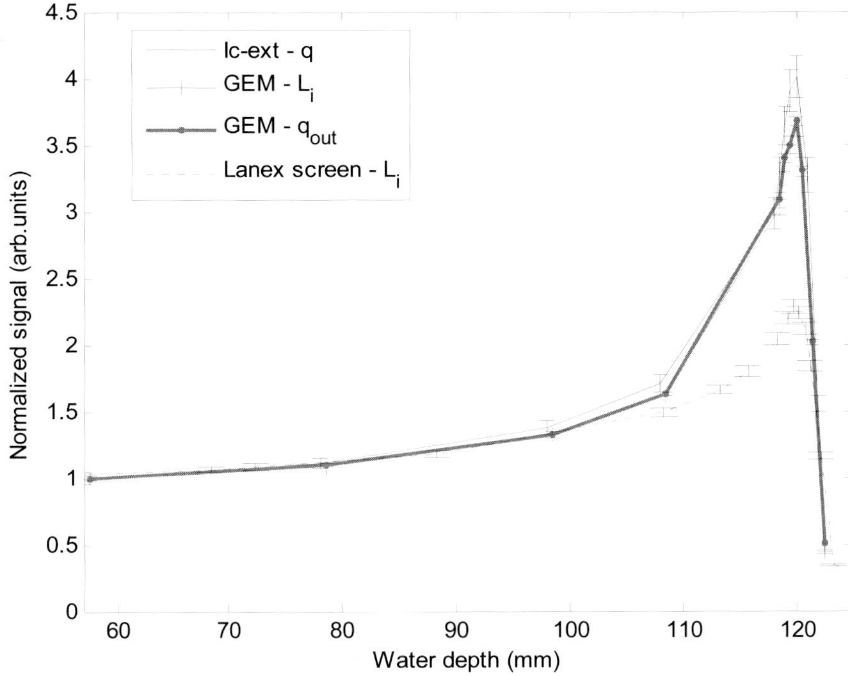

Figure 8.7 Comparison among Bragg curves measured with the reference ionization chamber, *Ic-ext*, the scintillating GEM detector and, the Lanex screen. All curves are normalized to one at the minimum water phantom thickness of 57 mm.

	Peak to plateau ratio
Ic-ext – q	4.00 ± 0.16
GEM – L_i	3.64 ± 0.08
GEM - q_{out}	3.67 ± 0.12
Lanex screen – L_i	2.28 ± 0.05

Table 8.1 Peak to plateau ratios of the Bragg curves represented in Figure 8.7.

This signal underestimation is much smaller than the one measured with the Lanex screen of about 43 %. To a lesser extent, the same was observed in an alpha particle beam [11]. A detailed analysis of the Lanex signal underestimation can be found in [8].

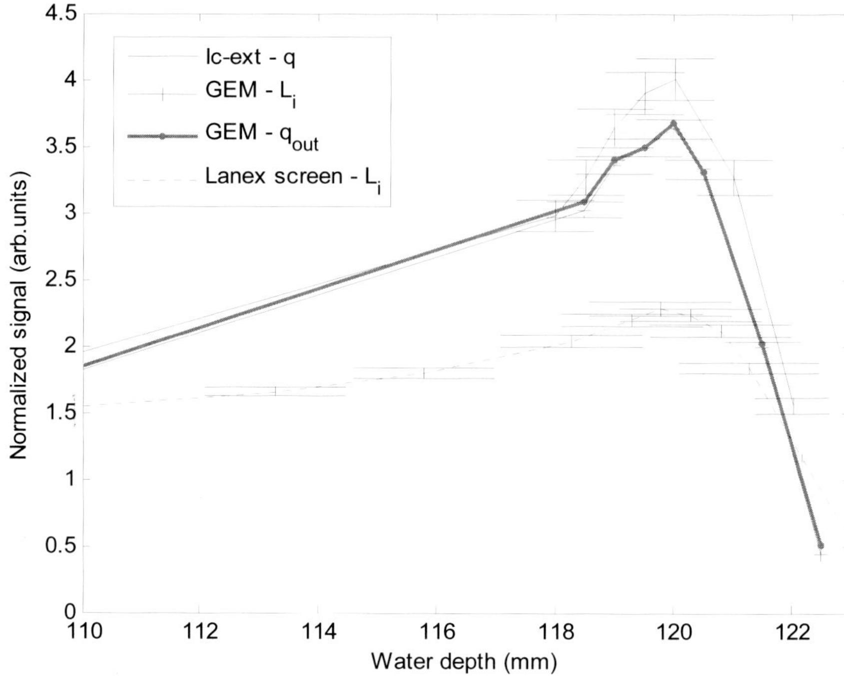

Figure 8.8 Enlargement of the Bragg peak region for the Bragg curves shown in Figure 8.7.

Part of the 9 % scintillating GEM detector signal underestimation, can be explained by taking into account the stopping power difference between Ar/CF_4 and air, the filling gas of *Ic-ext*. Measurements in [6] show that the Bragg peak to *plateau* signal underestimation of an Ar filled parallel plate ionization chamber is 3.5% with respect to when it is filled with air or N_2. These measurements were performed on the same beam line as we did our experiments with a 270 MeV/u beam.

L_i and q_{out} present the same signal underestimation within the uncertainties. Therefore, the cause of the remaining signal underestimation must be related to the charge creation process, which is responsible for both detector signals, and /or recombination in the *drift gap*. However, the latter can be excluded because carbon ion LET recombination effects were not observed for an Ar filled parallel plate ionization chamber as represented in Figure 3 of [23].

When a carbon ion interacts with the gas present in the *drift gap*, clouds of electrons and positive ions are formed around the trajectory of the primary carbon ion. The electron clouds drift away towards the GEMs quickly, while the ions move slower in the opposite direction. As can be seen in Table 8.2, the average distance D_{e-} of about 39 mm between the drifting electron clouds created by two different carbon ion tracks in the *drift gap* is quite big compared to the GEM hole pitch dimension of 90 μm.

Dose rate (Gy/s)	Carbon ion flux ($cm^{-2}s^{-1}$)	D_{e-} (mm)
0.03	$1.3 \cdot 10^6$	39

Table 8.2 Carbon ion dose rate, flux and average distance D_{e-} between the drifting electron clouds created by different carbon ion tracks in the scintillating GEM detector *drift gap*.

D_{e-} has been calculated taking into account the carbon ion flux and the electron drift time (0.05 μs in 90/10 Ar/CF_4 [24]). If the transversal diffusion coefficient (roughly 400 $\mu m/\sqrt{cm}$ in 95/05 Ar/CF_4 [25]) is also taken into account, one drifting electron cloud extends over about six GEM holes. However, it is still quite far from the next drifting electron cloud. Consequently, in one GEM hole there will be at most electrons created by a single primary carbon ion track. So, we think that the signal underestimation can not be attributed to too high a number of electrons clouds per GEM hole either. The average distances between ion clouds, not reported in Table 8.2, was found to be also larger than the GEM hole pitch taking also into account the diffusion coefficient. Therefore, space charge effects are not expected to influence the detector output and so to affect the signal at the Bragg peak depth.

A "saturation" of the charge creation process due to too high a number of multiplied electrons per GEM hole can not be excluded. In fact, the total number of electrons created by a primary carbon ion track and multiplied in the scintillating GEM detector, at the minimum water phantom thickness of the depth dose curve, is roughly $1.6 \cdot 10^6$ per cloud. At the Bragg peak depth, the total number of electrons per cloud is about 4 times higher, $\sim 6 \cdot 10^6$ electrons. These numbers are quite close to the phenomenological limit for gas multiplication before breakdown (*Raether limit*) [26] that according to literature for GEM based detectors is $\sim 10^7$-10^8 electrons per hole [27]. Therefore, a "non linearity", or "saturation" of the charge creation process could take place, especially near and at the Bragg peak depth, affecting the detector working conditions.

8.5 Conclusions

We have developed a 2D dosimetry system based on a scintillating gas detector, equipped with two cascaded GEMs in an Ar/CF_4 mixture. The photons emitted by the Ar/CF_4 electron-excited molecules, during the gas multiplication process, are detected by a CCD camera. Simultaneously to the light signal, GEMs currents are also measured for a better understanding of the detector operation. In this paper, we presented a first investigation of the properties of the scintillating GEM detector irradiated with a ^{12}C ion beam used for clinical radiotherapy treatment.

We have found that the output charge follows very well the beam intensity variations among spills and therefore also the time structure of the pulsed beam. The scintillating GEM detector light intensity at the minimum water depth thickness is 3 times brighter than the Lanex screen signal. For a typical beam spot, the scintillating GEM detector spatial response is ≤ 3.3 mm. The intrinsic GEM detector spatial resolution is expected to be at the sub mm level because of the low Ar/CF_4 diffusion coefficient. Moreover, the combination of the high degree of granularity of the GEM holes and of the CCD camera allows imaging with high resolution. However, the scintillating GEM detector has a slightly larger width at one tenth of the maximum with respect to that of the film. This indicates the presence of tails in the spatial distribution.

The scintillating GEM detector signal at the Bragg peak depth is only 9 % smaller than that of the reference ionization chamber. This result is much better than the response of the scintillating screen, which is 43 % smaller. The small scintillating GEM detector signal underestimation can partly (~ 3.5 %) be explained by the stopping power difference between the Ar/CF_4, filling gas of the GEM detector, and air in the reference ionization chamber. Since the charge and light output present the same Bragg peak depth signal underestimation within the uncertainties, the remaining signal underestimation must be related to the charge creation process, which is responsible for both detector outputs creation. Recombination in the *drift gap* and too high a density of electrons entering a GEM hole can in principle be excluded as causes of Bragg peak depth signal underestimation. However, a "saturation" of the charge creation process due to too high a number of multiplied electrons per GEM hole, especially at the Bragg peak depth, can not be ruled out.

Concluding, the scintillating GEM detector is a promising device for verifying dose distributions in carbon ions beams. However, further detailed studies on the detector characteristics (e.g. light yield, response reproducibility, rate capabilities, charge multiplication process) in a carbon ion beam are needed before it could be used in clinical practice.

Acknowledgments

The authors would like to thank Dr. D.Schardt for his support and help while performing the measurements at GSI.
This work was supported by Stichting voor Fundamenteel Onderzoek der Materie (FOM).

8.6 References

[1] U.Ramm et al., Three-dimensional BANG^{TM} gel dosimetry in conformal carbon ion radiotherapy, *Phys. Med. Biol.* 45, 2000, N95.
[2] M.J.Maryansky et al., NMR relaxation enhancement in gels polymerized and cross-linked by ionizing radiation: A new approach to 3D dosimetry by MRI, *Magn. Reson. Imaging* 11, 1993, 253.
[3] L.E.Olsson et al., MR imaging of absorbed dose distribution for radiotherapy using ferrous solphate gels, *Phys.Med.Biol.* 35,1990, 1623.
[4] C.P.Karger et al., A system for three-dimensional dosimetric verification of treatment plans in intensity-modulated radiotherapy with heavy ions, *Med.Phys.* 26,1999, 2125.
[5] G.C.Bonazzola et al., Performances of a VLSI wide dynamic range current-to-frequency converter for strip ionization chambers, *Nucl. Instr. and Meth. A* 405, 1998, 111.
[6] C.Brusasco et al., A dosimetry system for fast measurement of 3D depth–dose profiles in charged-particle tumour therapy with scanning techniques, *Nucl.Instr. and Meth. B* 168, 2000, 578.
[7] A.J.Lomax et al., Intensity modulated proton therapy: a clinical example, *Med.Phys.* 28, 2001, 317.
[8] S.N.Boon et al., Fast 2D phantom dosimetry for scanning proton beams, *Med.Phys.*25 1998, 464.
[9] S. Safai et al, Development of an inorganic scintillating mixture for proton beam verification dosimetry, *Phys. Med. Biol.* 49, 2004, 4637.
[10] S.N.Boon et al., Performance of a fluorescent screen and CCD camera as a two-dimensional dosimetry system for dynamic treatment techniques, *Med.Phys.*27 (2000), 2198.
[11] E.Seravalli et al., First Results of a Scintillating GEM Detector for 2D dosimetry in an Alpha Beam, *IEEE Trans. Nucl. Sci.* 54, 2007, 1271.
[12] S. Fetal et al., Dose imaging in radiotherapy with an Ar–CF_4 filled scintillating GEM, *Nucl. Instr. and Meth., A* 513, 2003, 42.

[13] F. Sauli, "GEM: A new concept for electron amplification in gas detectors," *Nucl. Instr. and Meth. A* 386, 1997,531.
[14] M. M. F. R. Fraga et al., "The GEM scintillation in He–CF_4, Ar–CF_4, Ar–TEA and Xe–TEA mixtures," *Nucl. Instr. and Meth., A* 504, 88, 2003.
[15] D.Schardt, Tumor Therapy with high-energy carbon ion beams, *Nuclear Physics A*, 787, 2007, 633.
[16] U.Weber, G.Kraft, Design and construction of a ripple filter for a smoothed depth dose distribution in conformal particle therapy, *Phys. Med. Biol.* 44, 2765.
[17] P. van Luijk et al., Stopping-power dependence of the light yield in a fluorescent-screen dosimeter, internal note.
[18] Matlab *medfilt2* routine, MATLAB® - The Language of Technical Computing, Mathworks Inc.,Natick MA, version 7.0.0.19920 (R14).
[19] F.Fraga, Performance of a tracking device based on the GEM scintillation, *IEEE Transactions on nuclear science*, 49, 2002, 281.
[20] CCD Image Sensor Noise Sources, application note, w w w . k o d a k . c o m / g o / i m a g e r s.
[21] Reibel et al., *CCD or CMOS camera noise characterization*, Eur. Phys. J. AP 21, 2003.
[22] B.Voss, private communication.
[23] T. Kanai, Initial recombination in a parallel-plate ionization chamber exposed to heavy ions, *Phys. Med. Biol.* 43, 1998, 3549.
[24] A. Peisert and F. Sauli, Drift and Diffusion of electrons in gases a compilation, *CERN* 84-08,1984.
[25]www.hep2.fzu.cz/ecfadesy/Talks/Tracking/Lepeltier_Vincent_ lepeltier_prague.ppt
[26] F.Sauli,Principles of operation of multiwire proportional and drift chambers, CERN report , 1982.
[27] Y. Ivaniouchenkov et al., Breakdown limit studies in high rate gaseous detectors, *Nucl. Instrum. and Meth. A* 422, 1999, 300.

Chapter 9

Conclusions and outlook

The main result of this work is the development and characterization of an innovative 2D dosimetry system for hadron beams based on a scintillating gas and Gas Electron Multipliers (GEMs) [1]. The detector has been characterized in an x ray beam, while its dosimetric properties were studied in a proton, alpha and clinical carbon ion beam.

As regards the detector operation, it has been found that the integrated light yield is linearly dependent on the total charge extracted from GEM_2 holes. This was expected since the photons are produced during the gas multiplication process.

With GEMs having 60 μm diameter double conical holes (small holes) a brighter signal and a higher electric output are measured than with wider (80 μm) holes. These facts are related to the electric field configuration along the hole axis. Under the same detector gain, the electric field strength for small holes is more intense in the hole centre compared to that for wider holes. Qualitatively, more intense electric field corresponds to higher ionization probability for Ar and for CF_4. Consequently, in a stronger electric field more electrons are created and more photons are emitted. The difference in light signal found with small and wider holes is bigger than the difference found in the output charge. In other words, for small holes the number of emitted photons per secondary electron is larger than that measured for big holes. This fact must be related to the excitation probability. If the excitation of Ar and CF_4 is more probable than their ionization, then more photons are produced with respect to the electrons. Further studies are needed to understand better this subject.

Varying the Ar/CF_4 gas mixture ratio, it was observed that for Ar + 8 % CF_4 the highest voltage across the GEMs, the highest electric output, q_{out}, and light output, L_i, values could be obtained.

It has been verified that the shape of the Ar/CF_4 emission spectrum in independent of (1) the voltages applied across the GEMs, (2) the x ray beam current, and (3) the GEM hole diameter. The ratio between the part of the spectrum attributed to CF_4 emission and that attributed to Ar changes when the Ar/CF_4 ratio is varied. The intensity of the Ar lines decreases when the CF_4 concentration is increased. In fact, CF_4 is a quencher gas that reduces the Ar photon emission. On the other hand, the height of the CF_4 visible

band is almost independent on the CF_4 concentration in the investigated range (from 4% to 20% CF_4). This may indicate the presence of an energy transfer between the Ar excited states and the dissociative electronic excited states of CF_4.
The magnitude of both L_i and q_{out} depends on the level of impurities inside the detector chamber. Reducing the gas flow rate or closing it affects the detector response more than varying the initial gas mixture purity. As the absence of leaks was established by means of the helium test[ccc], this indicates the presence of out gassing materials inside the chamber which "ages" the gas mixture. Efforts should be made in looking for less out gassing detector components, and in creating a cleaner gas handling system in order to have at least a constant level of impurities in the chamber.
It was noted that the detector response reaches an equilibrium value only a certain time after the irradiation has started. The time interval is about the same for both detector outputs. This so called "start up effect" influences the reproducibility of the scintillating GEM detector response. Therefore, in order to guarantee a more reproducible detector response, the "start up effect" should be further studied. In this context, the detector outputs must also be compensated for pressure variations. In chapter 5, an example of a pressure calibration curve was shown. For a more "user friendly" use of the dosimeter GEM detector, the compensation for pressure variations should be implemented in the data analysis routines or done "on-line" by means of, for example a feedback system.
As far as the dosimetric properties of the scintillating GEM detector are concerned, in a steady 150 MeV proton beam it was found that the detector response is linear with the delivered dose in the range 0.05 – 19 Gy within 1%. No dose rate effects were observed in the same beam in the range 1 – 16 Gy/min within 2%. Also at the Bragg peak depth, the detector response was found to be dose rate independent between 2 - 38 Gy/min within the experimental uncertainties. The scintillating GEM detector spatial resolution is in first approximation $\leq$ 1 mm. The intrinsic spatial resolution of the detector is expected to be at the sub mm level because of the low Ar/CF_4 diffusion coefficient. Moreover, the combination of the high degree of granularity of the GEM holes and of the CCD camera allows imaging with high resolution. In order to measure the intrinsic GEM detector spatial resolution, an experiment should be performed using a collimator with a slit

[ccc] The helium was flushed into the chamber using the same gas handling system used normally to flush the detector. Since no helium leakages were found (by means of a probe), we concluded that the detector was helium tight and therefore air tight. The presence of air was not ruled out from the system.

thinner than the expected spatial resolution. In general, the lower part of the GEM detector light intensity profiles is slightly wider than that of the Lanex screen. The origin of these longer tails in the spatial distribution has not been fully understood. It is not radiation beam related because also in a carbon ion beam, the FWTM of the GEM detector profile was found to be wider than that of a standard film optical density profile. These tails could be related to the presence of a residual background in the picture not compensated by the subtraction of the background picture. This residual background could be caused by light reflections and it is expected to be different for the GEM detector and Lanex screen. In the scintillating GEM detector, for example, the Al frames on which the GEMs are glued, the kapton, and the copper on the GEM surface are possible sources of reflections not present in the Lanex screen pictures.

The GEM detector light signal has a rise and fall time, of about 2 μs, which is faster than that of the scintillating screen signal (~ 1ms). The 2 μs rise time is probably related to the switching on time of the beam. Therefore, the scintillating GEM detector signal is expected to be even faster. A feasibility study was performed to check if the detector could work in a pulsed beam, in particular at dose rates typical of proton scanning beams. No major differences were observed in the outputs if 10 Gy at 15 Gy/min or 20 Gy at 240 Gy/min were delivered in one single pulse or in a large number of shorter pulses for the same CCD camera exposure time. Moreover, in a pulsed carbon ion beam (4s with a 50 % duty cycle), the electric output was found to follow very well the time structure of the beam.

No field size effects were observed in the range 120 – 3850 mm^2 within 2 %. The reproducibility of the detector output was found to be about 4-5 % over two days. Causes of this poor reproducibility are not yet well understood, and further studies should be performed on this subject especially in anticipation of a clinical usage of the detector.

The GEM detector response is not uniform over the GEM area. Inhomogeneities in the GEM detector emitted light intensity can be explained by a non uniform gain along the GEM surface [2,3]. The latter can be caused by hole shape variations during the manufacturing process, or non uniformly stretching of a GEM when it is glued onto the frame. Moreover, variations in the responsivity of the CCD camera pixels also play a role. The “non-flat” response can in principle be eliminated by calibrating, namely by taking an picture with a large field-shaping collimator every time new GEMs are mounted inside the detector chamber and used.

The GEM detector is radiation hard except for the glass exit window. The light transmission of the latter was found to be reduced after proton beam irradiation. This affects the reproducibility of the detector outputs and the response uniformity. A systematic study should be performed in order to quantify this radiation damage effect and its influence on the integrated light yield values. On the other hand, to get rid of this effect a quartz window instead of a glass window can be used.

The energy dependence of the scintillating GEM detector was studied not only in a 150 MeV proton beam, but also in a 360 MeV alpha particle beam, and a 250 MeV/u clinical carbon ion beam. Carbon ions, and to a lesser extent alpha particles, have a higher LET than protons. So, in this kind of beams the energy dependence of the detector response is emphasized. In all the three cases, the GEM detector output was compared along a Bragg curve to a parallel plate air filled ionization chamber (reference detector) and to the Lanex screen relative signal.

In the 360 MeV alpha beam (chapter 7), the measured underestimation of the scintillating GEM detector light and electric signals at the Bragg peak depth amounts to only 7% with respect to the reference ionization chamber relative signal. The scintillating screen signal is 34 % lower. 3% of the GEM detector signal underestimation is due to geometrical signal losses: at water layers thicker than 8 mm, the beam size was larger than the 70 mm diameter[ddd] entrance window. Therefore, part of alpha particles was lost compared to the reference ionization chamber. The remaining signal underestimation was not caused by dose rate effects. Part of it is due to the difference in the Ar/CF_4 – to - air stopping power ratio between the minimum water depth and the Bragg peak depth (considering a mono energetic alpha particle beam in the energy range of the measured Bragg curve, this difference is about 2%). So, summarizing the actual GEM detector signal underestimation in the alpha particle beam is only ~ 2 %.

In the 250 MeV/u ^{12}C ion beam (chapter 8), the scintillating GEM detector light and electric outputs at the Bragg peak depth[eee] are only 9% smaller than the reference ionization chamber relative signal. This result is very promising in comparison to the response of the scintillating Lanex screen, which is 43 % smaller. 3.5% of the scintillating GEM detector signal underestimation is attributed to the stopping power difference between Ar and air [4].

[ddd] The experiments in the alpha particle beam were performed with a GEM detector having slightly different layout than the one employed for the proton and carbon ion measurements. For details see chapter 7.

[eee] In this case the Bragg curves were normalized to one at 57 mm water depth because the data for 0 mm water depth could not be collected. The Bragg peak depth is at about 120 mm.

The Bragg peak depth signal underestimation measured in the 150 MeV proton beam is difficult to quantify because an anomalous difference was found between light and electric output peak to plateau ratio. The light signal underestimation at the Bragg peak depth is only 6 % with respect to an air filled ionization chamber. However, the electric output "quenching" is about 6-8 % higher. The signal underestimation of the detector outputs at the Bragg peak depth was expected to be the same because L_i was found to vary linearly with q_{out} at the plateau as well as at the Bragg peak depth. Moreover, in Bragg curve measurements performed in the alpha particle beam (chapter 7) and ^{12}C ion beam (chapter 8), q_{out} and L_i peak to plateau ratios were found to be comparable within 1%. From the available data, it has not been possible to explain this unexpected difference. Therefore, this experiment should be repeated. In addition, it was found that L_i peak to plateau ratio decreases for region of interest radii bigger than the reference ionization chamber sensitive area at the GEM_2 location. The same was also observed for the carbon ion data, where the peak to plateau ratio decreased for radii bigger than the GEM dimensions[fff]. Although in both cases, the decrease for "big" radii is within the experimental uncertainties, this phenomenon was not expected because the peak to plateau ratio should be independent of the region of interest, if the latter is chosen large enough to include the whole illuminated area. A possible cause of this peak to plateau ratio decrease can be related to the presence of a residual background in the pictures. As already mentioned, the latter could be caused by reflected light. The presence of reflections in the setup should be studied in detail in order to understand possible causes of light- scattering.

Concluding, the energy dependence of the response of the scintillating GEM detector in high LET beams is smaller than that of the Lanex screen. Moreover, the scintillating GEM detector output is linear with the dose and the field size. No dose rate effects were observed in the investigated ranges. The detector allows on line measurements with spatial resolution of the order to 0.5 mm and its time response is fast. Therefore, the scintillating GEM detector is a promising device for verifying dose distributions in high LET beams, e.g. for hadron therapy treatment plans composed by several beams of different energies.

However, a small signal underestimation with respect to a reference ionization chamber is still present. In order to eliminate the signal

[fff] For the alpha particle data nothing can be said because they were affected by geometrical signal losses and therefore the PPR was constant for radii bigger than those corresponding to the 70 mm entrance window diameter.

underestimation due to the stopping power difference between Ar/CF_4 and air or water, a better tissue equivalent scintillating gas mixture should be employed.
A theoretical study, for example by means of simulations, should be performed in order to understand the cause of the GEM detector signal underestimation, even though it is small. The fact that the light and electric output show the same signal underestimation at the Bragg peak depth within the experimental uncertainties (in alpha particle and carbon ion beam), indicates that the cause of this signal underestimation must be related to the charge creation process, which is responsible for both detector signals. "Too high" a number of drifting electron clouds (see below), or "too high" a number of multiplied electrons per GEM hole can affect the charge creation process.
When a hadron particle interacts with the gas present in the drift gap, clouds of electrons and positive ions are produced around the trajectory of the primary particle. The electron clouds drift towards the GEMs quickly, while the ions move slower in the opposite direction. This process is schematically represented in Figure 9.1.

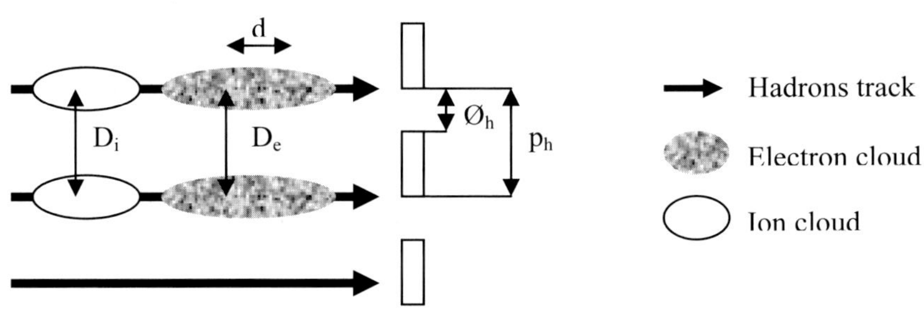

Figure 9.1 When an hadron particle (black arrow) interacts with the gas in the drift gap of the detector, it produces positive ions (white oval) and electron (gray oval) clouds around the trajectory of the primary particle. D_e is the average distance between the drifting electron clouds, while D_i the average distance between the ion clouds. d is the distance between electrons along the original hadrons track; $Ø_h$ GEM hole diameter (60 μm); p_h pitch of GEM holes (90 μm) [ggg].

[ggg] GEMs with small holes are considered.

In Table 9.1, D_e and D_i are shown for the average dose rates used in experiments with proton, alpha particle and carbon ions[hhh]. No dose rate effects were observed in the detector response within these dose rate ranges. D_e and D_i decrease when increasing the dose rate. Moreover, D_e and D_i of protons and alpha particles are about the same: on average, the distance between two different electron clouds is big compared to the GEM pitch of 90 μm. This means that there are not "too many" electron clouds per GEM hole. If the transversal diffusion coefficient of electrons in Ar/CF_4 is taken into account, then one electron cloud covers about 6 GEM holes. This distance (~ 540 μm) is still far from the next drifting electron cloud. The fact that the drifting electron clouds do not overlap is in agreement with the fact that no dose rate effects were observed experimentally in the dose rate ranges specified in Table 9.1. However, there is a chance that the ion clouds created by protons or alpha particles overlap. Consequently, space charge effects could occur.

Dose rate (Gy/min)	Flux ($1/(cm^2 \cdot s)$)	D_e (cm)	D_{ion} (cm)
Protons			
1.8	$3.45 \cdot 10^7$	0.76	0.020
30	$5.75 \cdot 10^8$	0.19	0.006
Alpha particles			
9.6	$3.3 \cdot 10^7$	0.78	0.020
54	$1.8 \cdot 10^8$	0.33	0.010
^{12}C ions			
1.8	$1.3 \cdot 10^6$	3.92	0.12

Table 9.1 Typical dose rates and fluxes in experiments performed with proton, alpha particle and carbon ion beams. Moreover, the average distance D_e between two drifting electron clouds and average D_i between two ion drifting clouds, calculated for the typical dose rates values of the three employed particle beams, are also listed.

For carbon ions, the average distances between electron and ion clouds are larger than the GEM hole pitch even if the diffusion is taken into account. Therefore, no effects on the detector operation are expected due to "too high" a density of electron clouds per GEM hole.

[hhh] D_e and D_i are calculated taking into account the hadrons flux and respectively the drifting time of electrons in Ar/CF_4, ~ 0.05 μs, and the drifting time of ions that is about 1000 time slower than the electron drifting time.

In Table 9.2, the total number of electrons multiplied by the two GEMs for the three used particle beams is shown at minimum water thickness (0-wd) and at the Bragg peak depth (Bp-wd). These numbers were calculated multiplying the number of electrons created in the drift gap, e^-, by the detector gain values of Table 5.1. At both water depths, the total number of multiplied electrons per electron cloud in carbon ions measurements is on average ten times higher than that during proton measurements with a detector gain that is half of the proton experiments gain (Table 5.1).

0-wd	d (μm)	e^-	Total e^-
Protons	40	80.2	$1.3 \cdot 10^5$
Alpha particles	7	458	$2.0 \cdot 10^5$
^{12}C ions	1	2092	$1.6 \cdot 10^6$
Bp-wd			
Protons	10	283	$5.0 \cdot 10^5$
Alpha particles	2	1786	$6.0 \cdot 10^5$
^{12}C ions	0.4	8368	$6 \cdot 10^6$

Table 9.2 Distance d between two electrons in the cloud; the 3rd column indicates the number of electrons created in the drift gap by the incoming beam; the 4th column indicates the total number of multiplied electrons per electron cloud. In order to calculate the latter, gain values of Table 6.1 were used.

This may indicate that the detector working conditions during carbon ion measurements were close to the Raether limit (section 3.2), which is ~ 10^7-10^8 electrons for GEM detectors according to literature [5]. Therefore, a "non linearity", or "saturation" of the charge multiplication process could take place, especially just before and at the Bragg peak depth, affecting the charge multiplication process.

Concluding, if in the carbon ion case the measured signal underestimation could be related to a "saturation" of the charge multiplication process, in the proton case no explanation is yet available for the observed lower signal at the Bragg peak depth compared to the reference ionization chamber.

In the alpha measurements (chapter 7), the integrated detector outputs were found to increase with dose rate in the investigated range. This increase was due to an initial instability ("overshoot") in the detector response when the beam is switched on. Experimentally, it was found that this "overshoot" was related to the magnitude of the voltage drop across the resistor value R_m (section 4.5) employed to measure the currents flowing on these GEMs surfaces. If the voltage drop across each GEM is smaller than 0.2 V, no

"overshoot" is measured. During proton and carbon ion measurements attention was paid choosing R_m values in order to satisfy the above condition. So, these experiments have not been affected by the presence of the "overshoot".
Two final remarks: the light signal magnitude of the GEM detector should be increased to eliminate, for example background related problems, and to be able to measure doses below the cGy level with the required accuracy. A simple way of increasing the light signal is to reduce the lens F-number, and/or to place the CCD camera closer to the GEM detector. Secondly, bigger GEMs, e.g. $40 \times 40\ cm^2$, should be tested because large detector area is one of the requirements for two dimensional dosimetry systems for hadron beams.

References

[1] F. Sauli, "GEM: A new concept for electron amplification in gas detectors," *Nucl. Instr. and Meth., A* 386, 1997,531.
[2] R.Bouclier et al., The Gas Electron Multiplier (GEM), *IEEE Transactions on Nuclear Science*, 44, 1997, 646.
[3] B.Yu et al., Study of GEM Characteristic for Application in Micro TPC, *IEEE Transactions on Nuclear Science*, 50, 2003, 836.
[4] C.Brusasco et al., A dosimetry system for fast measurement of 3D depth–dose profiles in charged-particle tumor therapy with scanning techniques, *Nucl. Instrum. and Meth., B* 168, 2000, 578.
[5] Y. Ivaniouchenkov et al., Breakdown limit studies in high rate gaseous detectors, *Nucl. Instrum. and Meth., A* 422, 1999, 300.

Summary

In radiotherapy, benefit for the patient can only be achieved if the treatment delivered to the patient is realized as planned. In particular, conformal treatments always bear the risk that an uncertainty in the delivered dose distribution may lead to an under dosage of the tumour, and/or over dosage outside the target volume. Therefore, it is mandatory to employ a quality assurance program that covers all steps of the treatment. This quality assurance program includes dosimetric verification of planned dose distributions, usually performed in water equivalent phantoms (that simulate the biological tissues). The aim of such dosimetric verifications is to check the proper transfer of treatment parameters and the correct execution of the plan at the treatment unit.

An ideal dosimeter for pre-treatment verification of dose distributions should be capable of acquiring real-time integrated tissue-equivalent signals with a fine spatial resolution ($\leq$ 1 mm) in three dimensions. Moreover, it should have a linear response over a large dynamic range in high intensity beams (~ 10^9 particles/($cm^2 \cdot s$)).

In hadrontherapy, one of the problems of currently used two- or three-dimensional dosimeters (e.g. scintillating screens, films) for pre-treatment verification of dose distributions is the energy dependence of their response in high LET (Linear Energy Transfer) radiation beams. The response of these detectors decreases for low particle energies due to saturation. As a consequence, these detectors underestimate the dose at the Bragg peak depth more than at the plateau of the depth dose curve. Correcting for this energy dependence is difficult because the composition of the particle field at each position in the irradiation volume and the corresponding detector responses must be known.

In an attempt to solve this limitation, we have developed and characterized a scintillating gas dosimetry system. The system is a follow up of the scintillating $Gd_2O_2S:Tb$ ("Lanex") screen setup [1,2]. With a gas as primary detection medium, in high LET radiation beams, a smaller energy dependence of the detector response is expected compared with that of a Lanex screen. Firstly, we expect that the light production process in a scintillating gas detector does not suffer from the quenching processes present in the Lanex screen. In fact, in the scintillating gas detector the photons are emitted by electron-excited gas molecules during the gas multiplication process. Secondly, the employed Ar/CF_4 scintillating gas mixture has better tissue equivalence (in terms of chemical composition) and a lower density than the scintillating screen.

The detector consists of two cascaded Gas Electron Multipliers (GEMs), mounted in an Ar/CF_4 scintillating gas mixture. A GEM is a copper clad thin kapton foil with a regular pattern of sub mm holes [3]. The primary electrons, created in the detector sensitive volume by the incoming beam, drift in an electric field towards the GEMs and undergo gas multiplication in the GEM holes. During this process, gas molecules are excited and they decay under fast light emission. In chapter 3, details about the operation principle of a scintillating gas detector based on GEMs and about the light emission can be found. The gas scintillation light is readout by a CCD camera. Since the amount of emitted light is proportional to the dose deposited by the incoming beam in the detector sensitive volume, the intensity distribution of the measured light spot is proportional to the 2D hadron dose distribution. By placing a water bellows phantom in front of the detector, with respect to the beam direction, and varying the water layer thickness in steps, from zero up to beyond the hadron range, a 3D dose distribution can be reconstructed. For a better understanding of the detector operation, simultaneous with the light signal also the currents flowing on the GEMs surfaces are measured.

In chapter 4, the measuring procedures are described. The sources of background present in a picture are classified. In particular, the background that takes into account the light coming from the scintillation of the glass exit window and of the Ar/CF_4 scintillation in absence of gas multiplication should be subtracted from the pictures. In fact, it was found that this light scales with the delivered dose, but it does not scale with the beam energy. In other words, it does not scale with the dose along a Bragg curve.

In chapter 5, the scintillating GEM detector characterization in an x ray beam is discussed. With GEMs having small holes (60 μm), a brighter light signal and a higher electric output are measured than with wider (80 μm) holes. The difference in light signal found with small and wider holes is bigger than the difference found in the output charge. It was observed that for Ar + 8 % CF_4 the highest voltage across the GEMs, the highest electric output and the highest light output could be obtained. It has been verified that the shape of the Ar/CF_4 emission spectrum is independent of (1) the voltages applied across the GEMs, (2) the x ray beam current, and (3) the GEM hole diameter. The ratio between the part of the spectrum attributed to

CF_4 emission and that attributed to Ar emission changes when the Ar/CF_4 ratio is varied. This can be explained by means of the model of Ar/CF_4 scintillation. The magnitude of both detector outputs depends on the level of impurities inside the detector chamber. Reducing the gas flow rate or closing it affects the detector response more than varying the initial gas mixture purity. It was noted that the detector response reaches an equilibrium value only a certain time after the irradiation has started. The time interval is about the same for both detector outputs. In this chapter, an example of compensation of the detector outputs for pressure variations has been illustrated.

The dosimetric properties of the scintillating GEM detector have been investigated in proton, alpha particle and ^{12}C beam. The aim of the proton experiments, reported in chapter 6, was to verify if the scintillating GEM detector fulfils the requirements for two-dimensional dosimetry systems to be used in proton beams. It was found that the scintillating GEM detector response is linear with the dose, and the field size; it does not show dose rate dependence; it can follow the time structure of a pulsed beam; it has a faster and brighter response than the Lanex screen and a spatial response ≤ 1 mm. However, some tails are present in the spatial distribution of the GEM detector light signal (full width at one tenth of the maximum is larger than that of the Lanex screen). The GEM detector response is not uniform over the GEM area. The response reproducibility is not within the requirements. The GEM detector is radiation hard except for the glass exit window.
The energy dependence was verified by means of a Bragg curve measurement. The light signal underestimation at the Bragg peak depth is only 6 % with respect to an air filled ionization chamber[iii]. However, the electric output signal underestimation is about 6-8 % higher. This difference between light and electric output peak to plateau ratio was not expected. The signal decrease found for the GEM detector is of the same order of magnitude of that measured for the scintillating Lanex screen (~ 14%).
As for the energy dependence of the scintillating GEM detector in an alpha particle beam (chapter 7) and a 250 MeV/u clinical carbon ion beam (chapter 8), it was found that the measured signal underestimation of both detector outputs at the Bragg peak depth is much smaller than the one presented by the scintillating Lanex screen. In the alpha particle beam, after correcting for geometrical signal losses, the signal underestimation at the

[iii] The response of the air filled ionization chamber is taken here as a reference.

Bragg peak depth is only 4%. 2% of this can be explained by taking into account the stopping power difference between Ar/CF_4 and air, the filling gas of the ionization chamber used as a reference.

	Peak to plateau ratio		
	Proton beam	Alpha particle beam	^{12}C ion beam
IC – q	3.53 ± 0.04	3.89 ± 0.04	4.00 ± 0.16
Lanex - L_i	3.03 ± 0.05	2.56 ± 0.04	2.28 ± 0.05
GEM - L_i	3.32 ± 0.09	3.72 ± 0.07	3.64 ± 0.08
GEM – q	3.11 ± 0.09	3.74 ± 0.07	3.67 ± 0.12

Table 1 Peak to plateau ratio of the Bragg curves measured by means of a reference ionization chamber, the scintillating Lanex screen, the scintillating GEM detector in respectively proton beam, alpha particle beam and ^{12}C ion beam. The peak to plateau ratio is defined as the ratio of a Bragg peak depth measurement and a measurement at the plateau of the Bragg curve.

In the 250 MeV/u ^{12}C ion beam, the scintillating GEM detector light and electric outputs at the Bragg peak depth are only 9% smaller than the reference ionization chamber relative signal. In this case, 3.5% of the scintillating GEM detector signal underestimation is attributed to the stopping power difference between Ar and air. This result is very promising in comparison to the response of the scintillating Lanex screen, which is 43 % smaller.
Therefore, the scintillating GEM detector appears to be a suitable device for verifying dose distributions especially in high LET beams, in which the currently position sensitive detectors present the problem of saturation. However, the light and electric output anomaly found in a proton beam has to be studied again.

Finally, in chapter 9 conclusions are given, including a summary of the main results and suggestions for directions for further research.

[1] S.N.Boon et al.,_Fast 2D phantom dosimetry for scanning proton beams, *Med.Phys.,* 25, 1998, 464.
[2] S.N.Boon et al.,_Performance of a fluorescent screen and CCD camera as a two-dimensional dosimetry system for dynamic treatment techniques, *Med.Phys,* .27, 2000, 2198.
[3] F. Sauli, "GEM: A new concept for electron amplification in gas detectors," *Nucl. Instr.and Meth. A,* 386, 1997,531.

Samenvatting

Radiotherapie is alleen behulpzaam voor de patiënt als de behandeling kan worden gerealiseerd als bedoeld. In het bijzonder bij conforme behandelingen waarbij het risico bestaat dat een onzekerheid bij de geleverde dosis distributie kan leiden tot een onder bestraling op de tumor en een over bestraling er omheen. Daarom is het belangrijk dat er een kwaliteitsprogramma wordt toegepast dat alle stappen van de behandeling na volgt. Een dergelijk kwaliteitsprogramma zal een verificatie van de voorbestemde dosis omvatten welk normaal uitgevoerd word met een in water geplaatste fantoom die het biologische weefsel simuleert. Het doel van deze verificatie is ter controle van de behandelingsparameters en de juistheid in de uitvoering van het behandeling plan.

Een ideale dosimeter voor het verifiëren van de dose distributie voordat de eigenlijke behandeling plaats neemt zal real-time weefsel-equivalente signalen moeten kunnen waarnemen bij een nauwkeurige ruimtelijke resolutie ($\leq$ 1 mm) in drie dimensies. Tevens belangrijk zal zijn een lineaire responsie over een groot dynamische bereik in hoge energetische stralen. (~ 10^9 particles/($cm^2 \cdot s$)).

Eén van de problemen van 2 en 3 dimensionale dosimeters (e.g. scintillating screens, films) voor verificaties bij hadron-therapie is de responsie afhankelijkheid in de hoge LET (Linear Energy Transfer) stralingsbereik. Responsies in deze detectoren verzwakken voor lag energetische deeltjes door een verzadigings effect. Dit heeft met als gevolg dat deze detectoren de dosis op de diepte van de *Brag peak* meer onderschatten dan op het continuüm van de dosis curve. Toepassen van een correctie voor deze energie afhankelijkheid is moeilijk omdat de compositie van het deeltjes veld op elke positie in het bestraalde volume bekend moet zijn met daarbij de overeenkomstige detector responsies.

Om deze limitatie op te losen hebben we een scintillerende gas dosimetry system ontwikkeld en gekarakteriseerd. Het systeem is een opvolger van het scintillerende Gd_2O_2S:Tb ("Lanex") scherm opstelling [1,2]. Met gas als primaire detectie medium is in hoge LET staling een lagere energie afhankelijkheid van de responsie te verwachten in tegenstelling met dat van een lanex scherm. Dit gebeurt omdat ten eerste, de licht productie process in een scintillerende gas verschilt met dat van het scherm. En ten tweede, het toegepaste Ar/CF_4 scintillating gas mensel geeft een beter weefsel equivalentie (in termen van chemische compositie) en heeft een veel lagere dichtheid dan dat van het scherm

Deze detector bestaat uit twee Gas Electron Multipiers (GEMs) in cascade geplaatst in een Ar/CF_4 scintillatie gas mengsel. Een gem is een kapton folie met een koperen laag waarin een constante patroon van sub-millimeter gaatjes zijn in aangebracht. De primaire electronen gegenereerd door de inkomende straal in de gevoelige volume van de detector bewegen in het electrische veld naar de GEMs waar zij een vermenigvuldiging zullen ondervinden in de gaatjes. Gedurende dit proces exciteren de gas moleculen en vervallen vervolgens met het uitzenden van fotonen.

Hoofdstuk 3 geeft een gedetailleerde beschrijving over de werking van de gas detector gebaseerd op GEMs. Sinds dat de hoeveelheid geëmitteerde licht proportioneel is aan de geleverde dosis van de inkomende straal; is ook de intensiteit distributie van het licht proportioneel aan de twee dimensionale dose distributie. Voor het beter begrip van de werking van de detector wordt tevens gelijktijdig met het lichtsignaal ook de electrische stromen van en naar de GEM oppervlakten gemeten. Met behulp van een water balg voor de detector kan de geleverde drie dimensionale dose distributie gereconstrueerd worden.

Hoofdstuk 4 beschrijft de meet methode. De bronnen van de achtergrond signalen worden tevens geclassificeerd. In het bijzonder het achtergrond signaal dat geproduceerd word in het glas van de uitree raam in de detector en dat van het Ar/CF_4 gas zonder de toepassing van de multiplicatie. Het is gebleken dat deze achtergrond signalen mee variëren met de geleverde dosis maar niet variëren met de geleverde energie van de straal. In andere woorden het achtergrond signaal varieert niet met de dosis langs een *Bragg curve*.

Hoofdstuk 5 bespreekt de karakteristiek van de scintillerende GEM detector in een rontgen straal. Gemeten is dat GEMs met kleiner gaten (60 μm) een feller licht signaal geven met een hogere electrische output dan die met bredere gaten (80 μm). Het verschil in het licht signaal bij smalle en brede gaten was groter dan bij het gevonden verschil bij de output lading. Er is geobserveerd dat Ar + 8 % CF_4 de hoogste voltage over de GEMs kan hebben met de meeste electrische output en het felste licht produceerde. Tevens is er geverifieerd dat het emissie spectrum van Ar/CF_4 gas onafhankelijk bleef met alle toegepaste voltages over de GEM, de intensiteit van de rontgen straal en met het formaat van de gaten in de GEM. De verhouding tussen het spectrum wat gegenereerd word door het CF_4 gas en dat wat gegenereerd word door het Ar gas veranderd wanneer de Ar/CF_4

mengverhouding wordt gevarieerd. Dit is te verklaren met een scintillatie model voor het Ar/CF_4.
De sterkte van de output van beide detectoren hangt erg af van de hoeveelheid onzuiverheden binnen de detector kamer. Het verminderen van de gas toevoer of het stoppen ervan heeft een grotere invloed op de responsie dan het variëren van de zuiverheid van het gas mengsel. Er is tevens gezien dat de responsie van de detector een evenwicht waarde bereikt enige tijd nadat de bestraling gestart is. De tijds interval voor deze is ongeveer gelijk voor beide detectoren. In dit hoofdstuk zal tevens een voorbeeld gegeven worden voor de compensaties omtrent de detector output in verband met druk variaties.

De dosimetrische eigenschappen van de scintillerende GEM detector zijn onderzocht in protonen, Alpha deeltjes en ^{12}C stralen. Zoals besproken in Hoofdstuk 6 is het doel van deze protonen experimenten om te verifiëren of de GEM detector geschikt is voor twee dimensionale proton dosimetrie systemen. Gevonden is dat deze detector lineair is met de dose en veld grootte; geen dose rate afhankelijkheid heeft; volgt het tijds structuur van een ge-pulste straal; sneller en fellere responsie geeft dan Lanex scherm en een ruimtelijke resolutie heeft dat kleiner is dan 1 mm. Doch is er een anomalie te zien in de ruimtelijke resolutie van de GEM detector licht signaal (de full-width bij een tiende van de maximum is groter dan bij het lanex scherm.). De GEM detector responsie is niet uniform over het gehele oppervlakte. De reproduceerbaarheid is niet binnen de vereiste eigenschappen. De GEM detector is radioactief bestendig behalve het glas van het uitree raam. De energie afhankelijkheid is gemeten met behulp van een *Bragg curve* meeting. Het licht signaal gaf een onderschatting van de *Bragg-piek-diepte* van 6% in vergelijking met een lucht gevulde ionisatie kamer[jjj]. Hoewel, het electrische output signaal onderschatting was in tegenstelling ongeveer 7-9 % groter. Dit verschil tussen licht en electrische output piek-tot-plateau verhouding was onverwacht.

Bij de energie afhankelijkheid van de scintillerende GEM detector in een Alpha deeltjes straal (hoofdstuk 7) en een 250 MeV/u klinische carbon ion straal (hoofstuk 8), is gevonden dat output onderschatting bij beide metingen op de *Bragg-piek-diepte* veel kleiner is dan bij een scintillerende Lanex scherm. In de Alpha straal, na correctie van geometrische verliezen, is de signaal onderschatting bij de *Bragg-piek-diepte* maar 4%, Hiervan kan

[jjj] De responsie van de lucht gevulde ionisatie kamer is hier genomen als een referentie.

2% verklaart worden aan de hand van het verschil in 'Stopping Power' tussen het Ar/CF_4 mengsel in de GEM detector en de lucht wat gebruikt is in de referentie ionisatie kamer

	Peak to plateau ratio		
	Proton beam	Alpha particle beam	^{12}C ion beam
IC – q	3.53 ± 0.04	3.89 ± 0.04	4.00 ± 0.16
Lanex - L_i	3.03 ± 0.05	2.56 ± 0.04	2.28 ± 0.05
GEM - L_i	3.32 ± 0.09	3.72 ± 0.07	3.64 ± 0.08
GEM – q	3.11 ± 0.09	3.74 ± 0.07	3.67 ± 0.12

Table 1 Piek tot plateau verhouding van de *Bragg curve* gemeten met behulp van een referentie ionisatie kamer, het scintillerende Lanex scherm, de scinitillerend GEM detector in respectievelijke proton, Alpha en ^{12}C ionen stralen. De piek tot plateau ratio is gedefinieerd als de verhouding tussen een meeting op *Bragg-piek-diepte* en een meeting op het *continuüm* van de *Bragg curve*.

In de 250 MeV/u ^{12}C ionen straal is de GEM detectors licht en electrische output op de *Bragg-piek-diepte* 9% kleiner dan de referentie ionisatie kamer. In dit geval is er een 3.5% onderschatting van het signaal te wijten aan het verschil van de 'stopping power' tussen Ar en Lucht. Dit resultaat is veelbelovend in vergelijking met de responsie van het Lanex scherm welk 75% lager bedraagt. Het is dus daarom dat het scintillerende GEM detector erg geschikt blijkt te zijn voor het verifiëren van dose distributie bij hoge LET stralen. Dit in het bijzonder omdat de huidige positie gevoelige detectoren problemen krijgen met de verzadiging. Doch blijft hier voor het moment een noodzaak voor verdere studie naar de anomalie van het verschil in de licht en electrische output in de proton stralen.

In Hoofdstuk 9 zijn de conclusies weergegeven met daarbij een samenvatting van de belangrijke resultaten en aansturing suggesties voor verder onderzoek.

[1] S.N.Boon et al.,_Fast 2D phantom dosimetry for scanning proton beams, *Med.Phys.*, 25, 1998, 464.
[2] S.N.Boon et al.,_Performance of a fluorescent screen and CCD camera as a two-dimensional dosimetry system for dynamic treatment techniques, *Med.Phys,* .27, 2000, 2198.
[3] F. Sauli, "GEM: A new concept for electron amplification in gas detectors," *Nucl. Instr.and Meth. A,* 386, 1997,531.

Acknowledgments

Well…. a lot of people contributed to this adventure that has started four years ago and they deserve to be thanked.

I am grateful to Carel van Eijk, my promoter, for his always fruitful contributions to my research and his positive approach towards problems. Thanks Carel for having read carefully my manuscript and for having discussed about it.

Thanks to Rob, my daily supervisor, for having pushed me in learning Dutch and having "forced" me in one way or another to learn rowing.

Marco Schippers, it is a pity we could not work closer. Your enthusiasm gave me strength to continue rowing when I was stranded. Thanks for your availability and for the sometimes animated but always stimulating discussions we had.

Martijn, I will not forget all the tips you gave me from the "sales world" and all the practical/technical ones. Thanks for your moral support and for having transform some difficult moments in a success.

I am grateful to Jan for having always answered my questions about electronics and not. Thanks for having shared the night shifts at KVI, and having hold out the tiredness when I was saying "Let's perform another experiment!" Thanks for having cared for.

Folkert, thanks for your constant availability and kindness. Thanks for having written the acquisition program that allowed us to employ efficiently our very short beam time at GSI.

I am very grateful to the guys of the workshop, in particular Renè and Kees, for having always done their best to be on time with deliveries; for having constructed the GEM detector; for having collaborated and helped when I was without technical support.

Thanks to Renè Gommers and Jeroen for the discussions we had about materials and the project of the detector.

Johan, thanks for having made possible the emission spectrum measurements and taught me how to play "klaverjas".

I would like to thank the operators of KVI, prof. Sytze Brandenburg and Peter van Luijk for their help during the beam time.

dr. Bernd Voss thanks for having helped us during the preparation of the measurements and during their execution at GSI.

Thanks to dr. M. Fraga and prof. F. Fraga for the interesting and stimulating discussions we had during my visit in Coimbra and for your availability.

The women of the group: Josè, with who I have almost shared an office in the last years. Thanks for being so proud of being a woman, for some discussions we had about life, for having helped to practice and so to improve my Dutch.
Candice thanks for always being there when a shoulder is needed or just for a chat. I really appreciated the times you brought me signs of spring from you garden and made me taste some delicatessen you prepared. Thanks for having cared for!
Romèe and Alice for the chats we had here and there, and during lunch breaks.
Together with Marnix and Andreas we shared the feelings of being stranded and we helped each other to refloat. Thanks for having listen to my worries, for the encouragement words and for the nice discussions we had not only about science. Marnix, a special thanks for having translated my propositions in Dutch, and for having commented on the English ones!
Thanks to Martijn Clarijs for having helped me with the "mfit" and to Danang for having taught me some tricks during the badminton sessions.
I am grateful to Peter for some suggestions he gave me about writing a dissertation.
Thanks Adrie for having read part of my manuscript.
Diane, Susana, Thea, Herman, Stefan, Alister, Jakob, Winicjusz, Pjotr, Dennis, Victor, Jan, Hans, Maijd thank you for having contributed in some way or another to my staying in the group.
I must not forget the Radiation Chemistry group for having allowed me to use their microwave oven for warming up my lunches.
Grazie ad Andrea, vecchio compagno di ufficio, per aver scritto i programmi di analisi delle figure. Da quando te ne sei andato, mi sono mancate davvero le nostre discussioni scientifiche e l'imparare il significato e la pronuncia della parola inglese del giorno.
Mirko, grazie per avermi fatto capire che cosa significa diventare genitore e aver condiviso momenti di "italianeità".
Jeffrey thanks for being so welcoming, for having shared good and less good moments and having translated the summary in Dutch.
Greg since you left, I have missed our daily chats together with a cup of coffee. Thanks for your advises and words of encouragement, they helped to reach this point. I have not spent yet the 5 euro cent you gave me to keep the project going…I'll invest them in the near future!
Grazie infinite Davide prima di tutto per avermi letto attentamente la tesi. E poi per aver condiviso momenti positivi e non del fare ricerca, per essere sempre stato disponibile al momento del bisogno.

Thanks to Tai and Aura for having accepted to be part of my team. For being such good friends in the way I mean friendship! For having made my staying here brighter!
To Audrey, Caroline, Christophe, Fatemeh, Esther thank you for having shared moments together.
I am grateful to the clouds that let the sun shine during almost the whole month of February making my writing days happier!
Le ultime righe le dedico a delle persone speciali senza le quali arrivare a questo punto non sarebbe stato possibile. Avrei dovuto mettervi in cima alla lista…I miei tifosi in Italia: mame, papi, Franci, Anna. Grazie per aver creduto in quello che stavo facendo e avermi dato coraggio. Anna, un grazie di cuore per avermi aiutato con le figure e la copertina.
E Ettore, il mio tifoso in Olanda. Grazie per avermi spinto a lasciare tutto e a partire… si è rivelata un'esperienza grandiosa. Grazie per avermi aiutata a resister durante i momenti più neri e per averli condivisi assieme a me.
Tutta l'energia positiva che vi succhiato cari tifosi… la potete ritrovare in queste pagine.

Wow…now it is really over…actually it is a new beginning, isn't it? ☺

Curriculum vitae

Enrica Seravalli was born on 15 May 1978 in San Daniele del Friuli, Italy. She attended high school in Udine where she received education in both science and humanities. In 2003, she obtained her master degree in Physics at the University of Padova with a thesis on "Microdosimetric characterization of a Boron Neutron Capture Therapy radiation field". She performed her graduation research at the Legnaro Laboratories of the National Institute of Nuclear Physics (INFN). Aim of the project was to characterize a Tissue Equivalent Proportional Counter for microdosimetric quality control of Boron Neutron Capture Therapy beams. In 2004, she started to work as a PhD at the University of Technology of Delft. This thesis is the result of her research that was supported by Stichting voor Fundamenteel Onderzoek der Materie (FOM).

List of publications

A Scintillating gas detector for 2D dose imaging in hadron beams.
E. Seravalli, M. De Boer, F. Geurink, J. Huizenga, R. Kreuger, J.M. Schippers, C.W.E van Eijk. Abstract in Radiotherapy and Oncology, Journal of the European Society for Therapeutic Radiology and Oncology, Vol. 84, Sep. 2007.

First Results of a Scintillating GEM Detector for 2-D Dosimetry in an Alpha Beam.
E. Seravalli, J. Hendrikse, J. Huizenga, R. Kreuger, J.M. Schippers, A. Simon, C.W.E. van Eijk, IEEE Transactions on Nuclear Science, Vol. 54, Issue 4, Part 3, Aug. 2007.

A scintillating triple GEM beam monitor for radiation therapy.
A. Simon, E. Seravalli, R. Kreuger, J. Hendrikse, E. Loeff, B.J.M. Heijmen, C.E.W.van Eijk, Nuclear Science Symposium Conference Record, 2005 IEEE, Vol. 5, Oct. 2005.

*BNCT Microdosimetry at the TAPIRO Reactor Thermal Column.*L. De Nardo, E. Seravalli, G. Rosi, J. Esposito, P. Colautti, V. Conte, G. Tornielli. Radiation Protection Dosimetry Vol. 110, 2004.

Conferences contributions

A Scintillating gas detector for 2D dose imaging in hadron beams.
E. Seravalli, M. De Boer, F. Geurink, J. Huizenga, R. Kreuger, J.M. Schippers, C.W.E van Eijk. 9th BIENNIAL ESTRO meeting on physics and radiation technology for clinical radiotherapy, September 8-13 2007, Barcelona, Spain.

A Scintillating Gas detector for 2D Dose Imaging in radiotherapeutic beams.
E. Seravalli, F. Geurink, J. Huizenga, R. Kreuger, J.M. Schippers, C.W.E van Eijk. 15th International conference on solid state dosimetry, July 8-13 2007, Delft, The Netherlands.

A Scintillating GEM detector for 2D Dose Imaging in HadronTherapy.
E.Seravalli, A.Simon, R.Kreuger, J.M.Schippers, C.W.E. van Eijk. PTCOG (Particle Therapy Co-Operative Group) 44, June 14-16, 2006 Zurich and PSI, Switzerland.

A Scintillating GEM Detector for Dose Imaging in Proton Therapy.
E.Seravalli, T.C. Delvigne, C.W.E. van Eijk, J.Hendrikse, R.Kreuger, J.M.Schippers, A.Simon. IEEE Nuclear Science Symposium and Medical Imaging Conference, 16-22 October 2004, Roma, Italy.